Big Data Analysis of Nanoscience Bibliometrics, Patent, and Funding Data (2000-2019)

Big Data Analysis of Nanoscience Bibliometrics, Patent, and Funding Data (2000-2019)

Edited by

Yuliang Zhao

Hongjun Xiao

Xingxing He

ELSEVIER

Elsevier
Radarweg 29, PO Box 211, 1000 AE Amsterdam, Netherlands
The Boulevard, Langford Lane, Kidlington, Oxford OX5 1GB, United Kingdom
50 Hampshire Street, 5th Floor, Cambridge, MA 02139, United States

Library of Congress Cataloging-in-Publication Data
A catalog record for this book is available from the Library of Congress

British Library Cataloguing-in-Publication Data
A catalogue record for this book is available from the British Library

ISBN: 978-0-323-91311-9

For information on all Elsevier publications visit our website at
https://www.elsevier.com/books-and-journals

Publisher: Glyn Jones
Editorial Project Manager: Naomi Robertson
Production Project Manager: Paul Prasad Chandramohan
Cover Designer: Alan Studholme

Typeset by TNQ Technologies

Contents

Editors and contributors

National Center for Nanoscience and Technology, China

Yuliang Zhao
Director, National Center for Nanoscience and Technology, China

Zhixiang Wei
Deputy Director, National Center for Nanoscience and Technology, China

Qing Dai
Director of Science & Technology Management, National Center for Nanoscience and Technology, China

Hongwei Dong
Nano Standardization Accreditation and Strategic Support Division, National Center for Nanoscience and Technology, China

Hongjun Xiao
International Affairs Office, National Center for Nanoscience and Technology, China

Shuxian Wu
Nano Standardization Accreditation and Strategic Support Division, National Center for Nanoscience and Technology, China

Elsevier and other contributors

Xingxing He
Analyst, Research Analytics and Data Service, Elsevier

Beverley Mitchell
Editor-in-chief, Science-Metrix, Elsevier

Thomas A. Collins
Data Scientist, Research Analytics and Data Service, Elsevier

Sarah Huggett
Head of Analytical Services APAC, Elsevier

John Van Orden
Strategy Associate, Elsevier

Christiane Barranguet
Director of Strategic Services, Elsevier

This book was created by Elsevier in collaboration with Professor Yuliang Zhao and members of the National Center for Nanoscience and Technology, China. Primary authors will include Xingxing He, Tom Thomas A., and Christiane Barranguet of Elsevier as well as Yuliang Zhao, Zhixiang Wei, Qing Dai, Hongjun Xiao, Hongwei Dong and Shuxian Wu from the National Center of Nanoscience Technology.

Executive summary

Over the past 2 decades, the number of nanoscience-related scientists and academic output has proliferated. The academic impact of nano-publications has steadily risen.

This book presents an evaluation of nanotechnology-related academic publications and patents from 2000 to 2019, focusing on nanotechnology's influence on basic scientific research and industry. The assessment shows research funding and international collaborations in nanoscience to be the driving forces of nanotechnology developments. The book's findings are based on literature citation, research analysis, funding, and patent data sources from Elsevier's Scopus database, SciVal, Funding Institutional, and PatentSight. The following are chapter highlights from this book.

Research efforts related to nanoscience rapidly increased over the past 2 decades.

(1) There were 1.42 million research publications on nanoscience, involving 2.21 million researchers.

The approximately 1.42 million nano-related academic publications accounted for 4.2% of publications worldwide between 2000 and 2019, according to Scopus. At the same time, 5.2% of total publication authors, about 2.21 million researchers,[1] published articles in the field of nanoscience.

(2) The scholarly output in nanoscience had an annual growth of 14.6% in the past 2 decades, which was 3.2 times the rate of global publications.

Globally, the number of nano-related publications (nano-publications) increased from 11,555 in 2000 to 153,455 in 2019,

[1] Researcher count method: This is based on the unique author IDs in Scopus, which are used as a proxy in calculating the number of researchers.

accounting for 1.1% and 6.2% of total annual academic publications, respectively. The compound annual growth rate (CAGR) for nano-publications was 14.6%, which was 3.2 times that of worldwide publications.

Meanwhile, the number of authors who published nano-related scholarly output increased from 32,591 in 2000 to 498,948 in 2019, accounting for 2.5% and 10.9%, respectively, of authors worldwide. The CAGR of nanoscience-publishing authors between 2000 and 2019 was 15.4%, which was 2.3 times that of global authors.

The field-weighted citation impact (FWCI) score of nano-publications in China and worldwide was rising, indicating a stronger academic impact in the field of nanotechnology in these areas.

(1) The FWCI of nano-publications was 1.6 times the global average.

For all global publications, the FWCI is benchmarked to 1. Between 2000 and 2019, the FWCI of nano-publications was 1.6. Nano-publications from China exhibited a significant increase in FWCI, shifting from 1.3 in 2000 to 1.9 in 2019, a growth rate of 43%. In 2019, China surpassed the United States' FWCI score in nanoscience.

(2) Among the top 1% highly cited publications, 11% were nanoscience-related, even though nano-publications accounted for only 4.2% of the global literature.

Over the past 2 decades, 11% of the top 1% highly cited publications were nano-publications. The total grew from 4.2% in 2000 to 13.6% in 2019, reflecting the essential contribution of nano-research in some of the most highly impactful studies.

As an interdisciplinary field, nanoscience served as a comprehensive platform for the integration of basic research, contributing to multiple fields and advancing scientific development.

(1) Nano-publications appeared most frequently in the subjects of materials science, chemistry, physics and astronomy, and energy.

Over the past 2 decades, the percentage of nano-publications has increased in multiple fields, especially in materials science. From 2000 to 2019, over 10% of studies in four research subjects were relevant to nanoscience. These subjects were materials science (20.7%), chemical engineering (17.7%), chemistry (16.3%), and physics and astronomy (12.8%). With 11% of its publications in nano-research, energy was added to the list for 2010–19.

(2) The fastest-growing subfields were in nanobiology.

In immunology and microbiology, the CAGR of nano-publications was 5.1 times that of all publications in the same field in 2000–19. This figure was 4.4 in the subfield of biochemistry, genetics, and molecular biology and 4.0 in the subfield of pharmacology, toxicology, and pharmaceutics. Despite the rapid growth, the volume of nano-publications in life sciences is still relatively small.

Nanoscience is closely related to the most prominent research topics.

(1) Nanoscience has covered some of the hottest issues in science.

Of the most prominent research topics[2] between 2015 and 2019, 89% had at least one publication related to nanoscience and 39% had strong connections to nanoscience: that is, at least 10% of their publications were nano-related. The data showed that nanoscience was tightly integrated with many emerging research fields.

(2) Among all fields, nanoscience had the closest connection with highly prominent topics in physical sciences.

Between 2015 and 2019, in the below 5 subjects - material science; physical and astronomy; chemistry; chemical engineering; energy; and pharmacology, toxicology and pharmaceutics - there were at over 42% of the most prominent topics strongly related to nanoscience, with at least 10% of the topic's publications being nano-related.

[2] The most prominent topics: These are topics whose topic Prominence score ranks among the top 1% in the world. See Chapter 2 and the Appendix for the topic significance score.

(3) Nanoscience focused on certain prominent topic clusters.

Between 2015 and 2019, nanoscience had the highest scholarly output in the highly prominent topics clusters[3] of solar cells, graphene, lithium batteries, plasmon metamaterials, biosensors, catalysts, semiconductor quantum dots, nanoparticles, and polymers.

Compared with China, the United States, Germany, Japan, and the United Kingdom had a higher academic—corporate collaboration rate[4] in nano-research, and patents frequently cited their results.

(1) In nanoscience, the United States, Japan, Germany, and the United Kingdom have a more extensive academic—corporate collaboration than does China.

Between 2015 and 2019, both China and the world had academic—corporate collaboration rates in nano-research that were slightly lower than the overall national and global averages, respectively. The United States, Japan, Germany, and the United Kingdom had higher rates than China, demonstrating more frequent collaboration between the academic and corporate sectors in these countries.

(2) The percentage of nano-publications cited by patents was higher than average.

Between 2015 and 2019, 1.04% of global nano-publications were cited at least once by patents filed under the largest five intellectual property offices.[5] This figure was 89% higher than the world average of 0.55% for all publications.

---

[3] Topic clusters with Prominence scores that rank in the top 5%.

[4] Academic—corporate collaboration rate: Literature copublished by authors from both academia and industry is referred to as academic—corporate collaborated publication. The ratio of academic—corporate collaborated publications to all publications is the academic—corporate collaboration rate.

[5] The largest five intellectual property offices: World Intellectual Property Organization, the US Patent and Trademark Office, European Patent Office, Japan Patent Office, and United Kingdom's Intellectual Property Office.

The five offices recorded an average of 10.4 citations per 1000 nano-publications. Of China, the United States, Germany, the United Kingdom, and Japan, the United States had the highest rate, at 23.7 citations per 1000 nano-publications, and China had 6.0.

(3) Nano-related patents from China have been proliferating, but China's patent competitiveness has room for improvement.

In the past 2 decades, there were about 690,000 nano-related patents worldwide, according to data from the patent analysis platform PatentSight. The percentage of nano-related patents grew from 0.9% in 2000 to 3.8% in 2019. Among those nano-patents, 58% came from China. Although the country has the largest number of nano-related patents, there is room for improvement in its competitiveness.

Among the projects funded by major global funders, the share of nano-related grants continued to increase.

(1) The share of nano-related awards was rising.

According to the funding database Funding Institutional, 132,220 awards were granted to nanoscience-related projects between 2009 and 2018. The number accounted for approximately 3.6% of all global awards in the database, and the CAGR was 3%. Although the total number of grants in that period remained the same, the proportion of nano-related projects increased from 3% in 2009 to 4% in 2018.

(2) Materials science received the highest share of nano-related awards.

Among the fields of science, materials science received the most grant awards for nano-research. Between 2009 and 2018, 29.4% of awarded grants were relevant to nanoscience in materials science, followed by 17.9% in physics and astronomy, and 14.8% in chemistry.

The degree of international collaboration in nanoscience was higher than the average for all research fields.

(1) The international collaboration rate in nanoscience was higher than all research fields combined.

Between 2010 and 2019, the international collaboration rate in nanoscience was 25%, with a total of 277,793 nano-publications published by collaborating authors from different countries and regions. The worldwide international collaboration rate over that period was 21%, indicating more frequent international cooperation in nanoscience.

(2) China has proven itself as a global ally in the nanoscience field.

China's international collaboration efforts in nanoscience have steadily risen, and at a rate much faster than those of other research disciplines in the country. Meanwhile, internationally collaborated nano-research with China as a partner exhibited a high academic impact: between 2010 and 2019, international nano-publication collaborations from China had an FWCI of 2.5, a higher score than those of other countries such as the United States (2.3), Japan (1.8), Germany (1.8), and the United Kingdom (1.9).

About the National Center for Nanoscience and Technology and Elsevier

National Center for Nanoscience and Technology

The National Center for Nanoscience and Technology, China (NCNST), established in December 2003, is co-founded by the Chinese Academy of Sciences (CAS) and the Ministry of Education as an institution dedicated to fundamental and applied research in the field of Nanoscience and technology, especially research with important potential applications. NCNST is operated under the supervision of the Governing Board and aims to become a world-class research center, as well as a public technological platform and young talent training center in the field, and to act as an important bridge for international academic exchange and collaboration.

Elsevier

This book has been prepared and published by Elsevier's Analytical Services, part of Elsevier's Research Intelligence portfolio of products and services. For over 140 years, researchers and healthcare professionals have trusted and relied upon Elsevier's portfolio of journals, including iconic publications such as The Lancet, Cell Press, and Gray's Anatomy. Today, Elsevier is a global leader in information and analytics, helping research communities advance science and improve health outcomes for the benefit of society by facilitating insights and critical decision-making for customers across the global research and health ecosystems.

Foreword

Nanoscience, a critical frontier field in science and technology in the 21st century, has long had an impact on economic and social development and thus has been a focus of global attention. It is a highly interdisciplinary field with applications across a wide range of sectors, from aviation to clinical solutions to the energy industry. With its numerous applications and interdisciplinary nature, understanding how research in the field of nanotechnology has evolved in recent years is essential to successfully solving the modern strategic and societal challenges faced by all nations.

Over the past 2 decades, the field of nanoscience has become increasingly accelerated, in part owing to improved collaboration across disciplines through the launch of dedicated programs, research centers, and other government initiatives around the world. In the United States, the National Nanotechnology Initiative was founded in 2000 to coordinate nano-related research and resources across 20 different federal departments. More recently, China's 14th Five-Year Plan outlines the importance of frontier areas in science and technology while acting on innovation-driven development strategies. Thus, it is unsurprising to find high-end smart materials such as shape memory alloys, self-healing materials, and nano-functional materials such as graphene and metamaterials among the 100 major projects of the 14th Five-Year Plan. The plan also identifies frontier nano-research as one of the National Key Research and Development Objectives to advance scientific exploration at the nanometer scale. As the importance of nanoscience is increasingly recognized, I believe more and more dedicated research organizations around the world will arise to nourish its growth.

Of course, much progress has already been made in nanoscience that needs to be celebrated. As this in-depth and comprehensive analytical book reveals, the subject has seen enormous growth in research output, with varied industry partners across many different sectors between 2000 and 2019. In fact, nano-related research has risen from only 1.1% of all global research in 2000 to an impressive 6.2% by 2019. Such progress merits highlighting in an accessible manner, for both experts and nonscientists alike, to demonstrate better the fundamental role of nanoscience in technology and everyday quality of life.

Looking forward, future progress will undoubtably rely on strengthening basic research in the field, coupled with an enhanced awareness and improved utility of the partnerships across industries that drive research. It is also critical to plan for best practices to transform those basic research results into applied technological products. This is one of the most relevant strategic goals in terms of the development of nanotechnology in China, as well as globally.

Big Data Analysis of Nanoscience Bibliometircs, Patent, and Funding Data aims to further an understanding of the development of nanoscience from five perspectives based on bibliometric analysis. It reviews the development of nano research over the past 20 years, revealing the impact of nanoscience on other research fields and clarifying the development of nano research from basic research to industry applications. It also summarizes key countries' nano research development strategy based on funding analysis and research focus analysis and anticipates upcoming frontier research in the nano field.

The book in its entirety provides an overall explanation of the current status and future development of nanoscience from a macro perspective, and provides extremely valuable data support and factual references for the realization of scientific and technological policy as well as major breakthroughs in the field of nanoscience.

It is my sincere hope that the field of nanoscience will advance with each passing day, and that the nano industry will continuously find new and ambitious ways to benefit our society. I believe that if we continue to watch the field's rapid pace of evolution closely, we will not be disappointed.

Dr. Chunli Bai
President, University of Chinese Academy of Sciences
Former president, Chinese Academy of Sciences

Introduction

THE FUTURE DEVELOPMENT OF SCIENCE WILL CONTINUE TO MAKE INROADS INTO THE MACRO AND MICRO WORLDS.

The invention of the scanning tunneling microscope in 1981 gave birth to nanoscience and nanotechnology, which aided scientists' exploration of the microscopic world between nanometers, a metric unit of length that describes a billionth of a meter. Nanoscience and nanotechnology refer to the research of quantum properties and interactions of substances at the nanoscale, such as atoms and molecules. These also seek to investigate the interdisciplinary sciences and technologies that leverage these characteristics. Through the lens of nanoscience and nanotechnology, humankind's understanding of the world extends, and new means to shape the world at the atomic and molecular level emerge. Nanotechnologists aim to produce products with specific functions based on nanoscale substances' novel physical, chemical, and biological properties[1].

By assessing and changing the world's future one micron at a time, the field of nanoscience and nanotechnology has been drastically expanding since the 20th century. As a young and dynamic research and application field, it is transforming the world as we know it, delivering revolutionary advancements to industries such as manufacturing and health care. Aided by their interdisciplinary, comprehensive, and fundamental characteristics, nanoscience and nanotechnology have become the driving force in science development.

Nanotechnology has also established its value in various economic sectors. Besides the nanomaterials field in the new material industry, nanotechnology applications in the areas of energy and environment, biomedicine, information devices, and green manufacturing have become increasingly prominent, with promising prospects.

[1] Bai, Chunli. (2005). Nanoscience & technology: Dream and reality. Proceedings of the 2004 China Nanotechnology Application Symposium.

Given the strategic significance of nanoscience and nanotechnology, it is essential to evaluate and predict their developmental trends. This book is based on the largest abstract and citation database of peer-reviewed literature, Elsevier's Scopus, along with the research evaluation platform SciVal. The funding and patent analysis data for the book are from Funding Institutional and PatentSight, respectively. **This book provides an evaluation of Nanoscience's scientific output, role, contribution, and impact** through bibliometric analysis, combined with big data indicators of nanoscience scientific results and patents from 2000 to 2019.

Please refer to the Appendix for the definition of nano-publications and the indicators used in the book.

CHAPTER

1 Scholarly output and academic impact of nanoscience

Key findings

4.2%
of global scholarly output were nano-publications (2000–19).

2,211,585
authors published nano-publications (2000–19).

1.6
was the field-weighted citation impact for nano-publications, which was 1.6 times the world average (2000–19).

28.8
citations per paper were received by nano publications, which is 48% more than the world average (2000–19).

11%
of output in the top 1% citations was nano-related (2000–19).

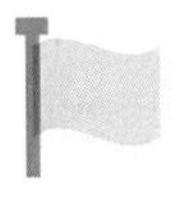

25%
of output in the top 1% citations in China was nano-related (2000–19).

The growth rate of nanoscience was several times the average for many key indicators, which reflects the full flourishing of nanoscience (Table 1.1).

Big Data Analysis of Nanoscience Bibliometrics, Patent, and Funding Data (2000–2019)
https://doi.org/10.1016/B978-0-323-91311-9.00001-3

Table 1.1 Global totals for nano-related authors, scholarly output, and output in top 1% highly cited publications (2000–19).

Nano-related	2000–19 overall number (share)	Number in 2000 (share)	Number in 2019 (share)	Growth rate of 2019 compared with 2000	CAGR of nano-related items (2000–19)	CAGR of nano-related items/CAGR of all fields' items
Author counts	2,211,585 (5.2%)	32,591 (2.5%)	498,948 (10.9%)	1,431%	15.4%	2.3
Scholarly output	1,418,496 (4.2%)	11,555 (1.1%)	153,455 (6.2%)	1,228%	14.6%	3.2
Top 1% highly cited publications	54,052 (11%)	585 (4.2%)	5393 (13.6%)	374%	8.5%	1.5

CAGR, *compound annual growth rate.*
From Scopus.

1.1 Overview of scholarly output and academic impact of nanoscience

This section provides an evaluation of global nano-related research output and its academic impact between 2000 and 2019. Assessment indicators include scholarly output, citation count, author count, field-weighted citation impact (FWCI), and publication output in the top 1% citations. Details about the definitions of these indicators and values can be found in Appendix A. Further information about the data is presented in Fig. 1.1.

Based on these data on nanoscience academic output and impact, further findings are presented subsequently.

(1) Nanoscience contributed significantly to global scientific research. The global share of nanoscience research output was relatively high: 4.2% of global research output was related to nanoscience. Approximately 1.4 million publications in nanoscience (nano-publications) had a total citation count about 40.9 million, contributing to 6.4% of all citations worldwide.

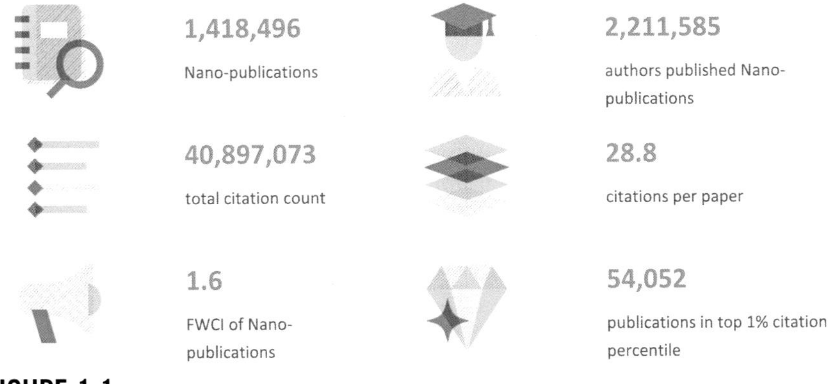

FIGURE 1.1

Overall nanoscience scholarly output and impact in the world (2000–19). *FWCI*, field-weighted citation impact.

From Scopus.

(2) The FWCI of nano-publications was higher than that of the world average. A citation refers to the source of information, such as academic literature, used in other research work. The citation count is one of the most widely recognized indicators for evaluating academic impact. However, assessing a publication's influence is challenging because of differences in citation practices across fields, publication years, and document types. To address the issue, Elsevier developed a standardized impact indicator, the FWCI, to appraise research academic impact of publications. This book includes relative indicators, such as citations per publication and highly cited publications, to present full academic impact of nanoscience.

Some overall statistics from 2000 to 2019 are:

- Whereas the global average FWCI was 1, the FWCI of nano-publications worldwide was 1.6, indicating that nano-publications' academic impact was 1.6 times the global average.
- Citations per paper for nano-publications totaled 28.8, 48% higher than the world average, which was 19.4.
- Of the top 1% highly cited publications, 11% were nano-related, for a total of 54,052 publications.

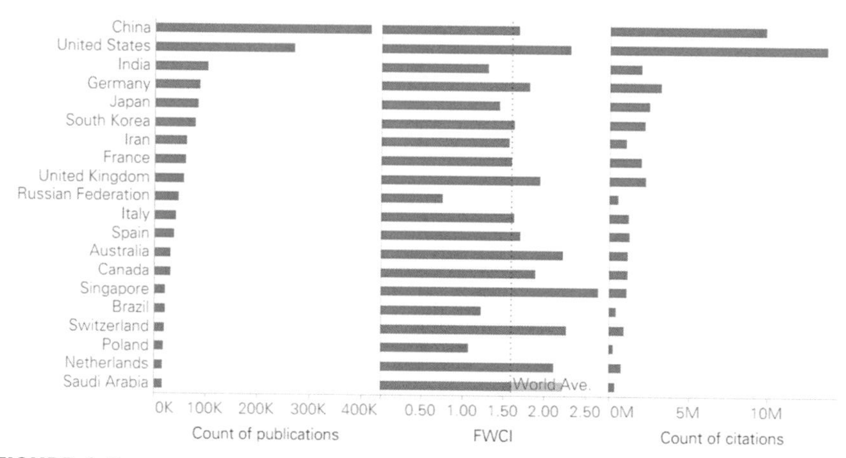

FIGURE 1.2

Top 20 countries by nano-publication output (2000–19). *Ave.*, Average; *FWCI*, field-weighted citation impact.

From Scopus.

(3) China ranked first for nano-publication output worldwide. Among the 20 countries with the highest scholarly outputs in nanoscience between 2000 and 2019, the top five were China, the United States, India, Germany, and Japan (Fig. 1.2). The five countries' nano-publications accounted for 29.4% (416,554 publications), 19% (269,747 publications), 7.3% (102,904 publications), 6.1% (87,164 publications), and 5.9% (54,079 publications) of global nano-publications, respectively.

Between 2000 and 2019, the top five among the 20 countries with the highest FWCI scores for nano-publications were Singapore (2.65), the United States (2.31), Switzerland (2.26), Australia (2.22), and the Netherlands (2.12). In the same period, the average world FWCI of nano-publications was 1.6. However, the output volumes of nano-publications in Switzerland, the Netherlands, and Singapore were lower than for other leading countries.

The top five countries by total citation count were the United States (13,762,200), China (9,854,878), Germany (3,222,746), Japan (2,498,856), and the United Kingdom (2,270,916). To provide a comprehensive comparison based on the volume and impact of academic output in nanoscience across countries between 2000 and 2019, the book selected these five countries as the key comparators for further analysis.

1.2 Trends in scholarly output in nanoscience

The number of nano researchers worldwide is rising

Research talent is a crucial element for science advancement. This section assesses current developments in nanoscience by analyzing publication authors. Refer to Appendix A for descriptions of the statistical methods used.

The number of authors who published nano-publications continued growing between 2000 and 2019 (Fig. 1.3), indicating a rise in scientists engaged in nanoscience-related work or integrating nanotechnology into their field. The total number of authors publishing nano-related research grew from 32,591 in 2000 to 498,948 in 2019, accounting for 2.5% and 10.9%, respectively, of the researcher population worldwide. Over the past 2 decades, the compound annual growth rate (CAGR) of authors in nanoscience was 15.4%, which was 2.3 times the global CAGR.

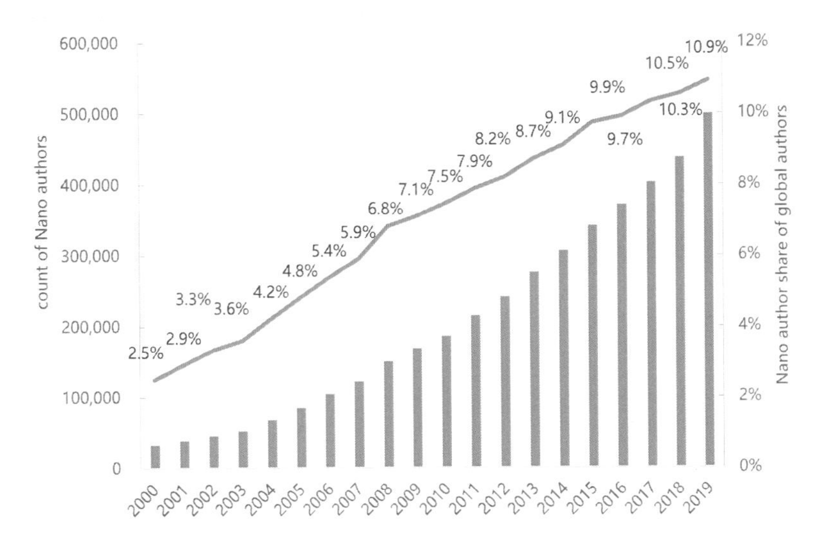

FIGURE 1.3

Number and global share of authors with nano-publications (2000–19).

From Scopus.

Nano-related academic output is rising

In this book, academic output is measured by the number of nano-publications. The volume of global nano-related output proliferated between 2000 and 2019, its share of worldwide research output rose as well. The number of nano-publications increased from 11,555 in 2000 to 153,455 in 2019, accounting for 1.1% and 6.2%, respectively, of global publications in each year (Fig. 1.4). Whereas the CAGR for global publications was 4.5%, the CAGR of nano-publications in the past 2 decades was 3.2 times higher, at 14.6%.

The nano-related academic output and impact of key countries (China, Germany, Japan, the United States, and the United Kingdom) are analyzed in this book. In general, the number of nano-publications in each key country grew steadily. The CAGR of nano-publications in these countries exceeded each country's average CAGR for all academic output (Fig. 1.5). The analysis shows the rapid development of nanoscience in these countries.

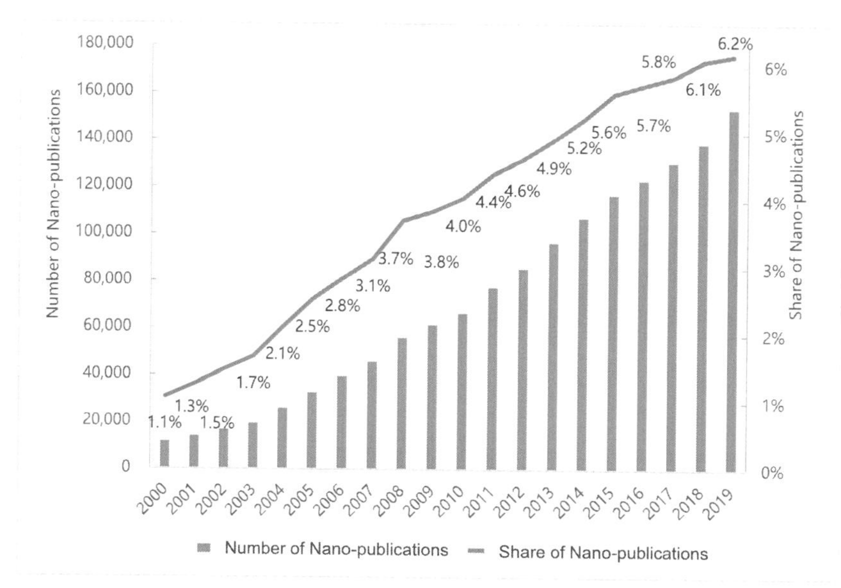

FIGURE 1.4

Number and global share of nano-publications (2000–19).

From Scopus.

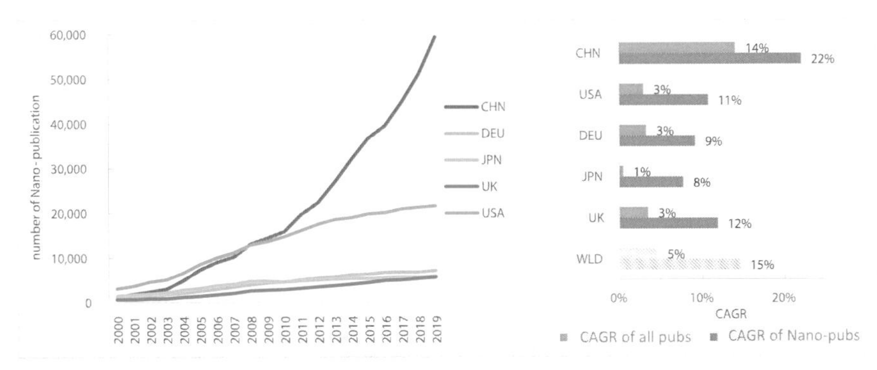

FIGURE 1.5

Scholarly output (left) and compound annual growth rate (CAGR) (right) of nano-publications in key countries (2000–19). *CHN*, China; *DEU*, Germany; *JPN*, Japan; *UK*, United Kingdom; *USA*, United States; *WLD*, world.

From Scopus.

China had the highest growth rate for the number of nano-publications among the comparators (Fig. 1.5), and its output accounted for the highest percentage of global nano-publications. China's nanoscience publications grew from 1,341 in 2000 to 59,349 in 2019, which accounted for 11.6% and 38.7% of nano-publications worldwide for each year. The CAGR of China-published nanoscience research was 22%, which was 1.6 times the nation's total academic output growth rate (CAGR = 14%). Over the same period, the amount of US-published nanoscience output increased from 3,115 in 2000 to 21,608 in 2019, accounting for 27% and 14.1% of nano-publications worldwide, with a CAGR of 12%. The growth rate in the field was four times the average of the growth rate for all publications in the United States (CAGR = 3%) in the same period.

The analysis showed that the percentage of nano-publications from China had been continuously growing (Fig. 1.6). The share of nano-publications also varied in each country (Fig. 1.7), the result of different focuses of high-yield research in each country. Statistics from the Scopus database for 2000–19 indicate that the biggest four research fields in China were engineering, materials science, physics and astronomy, and chemistry, which are subjects closely related to

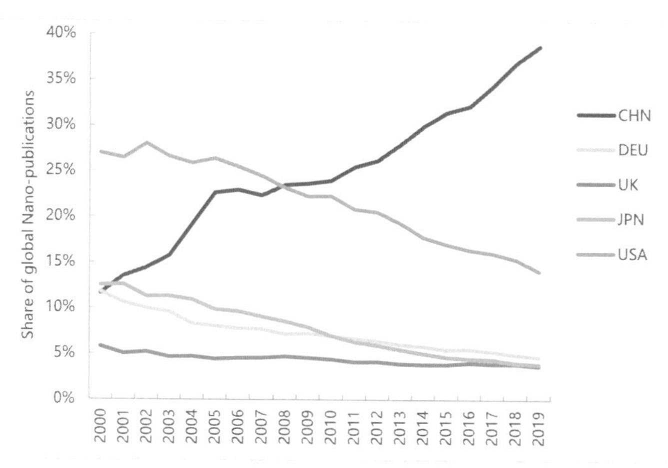

FIGURE 1.6

Share of global nano-publications per country (2000–19). *CHN*, China; *DEU*, Germany; *JPN*, Japan; *UK*, United Kingdom; *USA*, United States; *WLD*, world.

From Scopus.

FIGURE 1.7

Nano-publications' share of all publications per country (2000–19). *CHN*, China; *DEU*, Germany; *JPN*, Japan; *UK*, United Kingdom; *USA*, United States; *WLD*, world.

From Scopus.

nanoscience. In the United States, fields with the highest academic output were medicine and biochemistry, genetics, and molecular biology. To date, nano-related research in these fields is less than in engineering, materials science, physics and astronomy, and chemistry. Similar disciplinary differences were also existed in Germany, the United Kingdom, and Japan.

1.3 Trends in the academic impact of nanoscience

Nano-publications scored a higher field-weighted citation impact than the global average, and China's field-weighted citation impact of nano-publications had steadily increased

FWCI is an indicator of the citation impact of a publication. It is calculated by comparing the number of citations received by a publication with the number of citations expected to be received by a publication of the same document type, publication year, and subject. An FWCI of more than 1.00 indicates that the entity's publications have been cited more than would be expected based on the global average for similar publications. For example, an FWCI of 2.11 means the publications of the entity in question were cited 111% more than the world average. For further details about this indicator, please refer to Appendix A.

Between 2000 and 2019, the overall FWCI of nanoscience publications declined slightly. However, the FWCI scores increased slightly between 2015 and 2019. In addition, the figures were higher than the worldwide average for all publications (Fig. 1.8). The decline in the FWCI of global nano-publications resulted from the rapid growth in new nano-research, in which accumulated nano-publications outnumbered accumulated citations (Fig. 1.9). This drop in impact is a common occurrence when there is high growth of publication output. In the early development stages of nanoscience, in 2000–05, many high-profile publications that were considered to be classics were cited intensively, spiking the FWCI of nano-publications in the early days.

At the national level for China, the nano-publication FWCI increased from 1.3 in 2000 to 2.9 in 2019, achieving a 43% growth

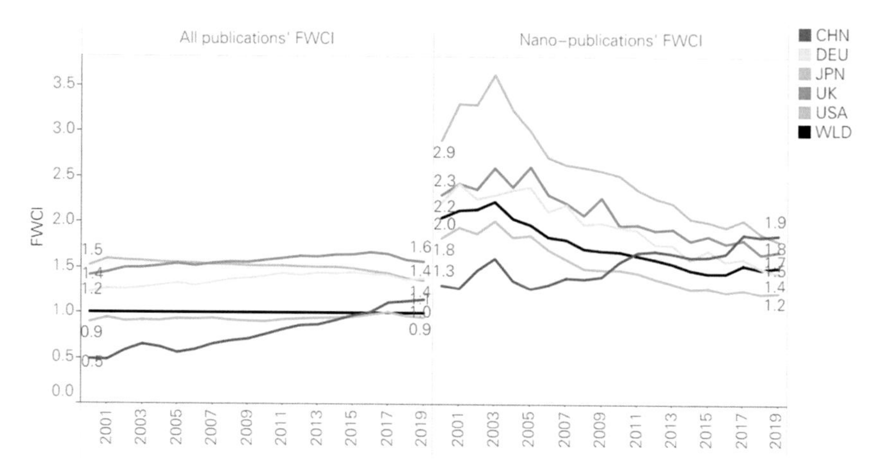

FIGURE 1.8

Trends in nano-publication field-weighted citation impact (FWCI) for the world and leading countries (2000–19). *CHN*, China; *DEU*, Germany; *JPN*, Japan; *UK*, United Kingdom; *USA*, United States; *WLD*, world.

From Scopus.

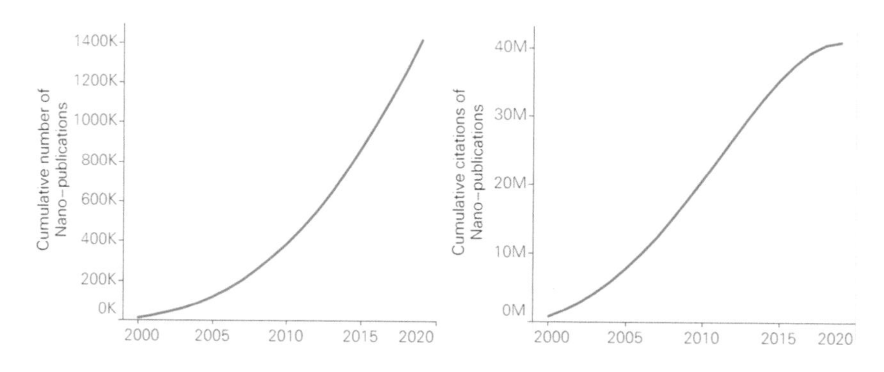

FIGURE 1.9

Trends in accumulated scholarly output and citation count of global nano-publications (2000–19).

From Scopus.

over the two decades. Nano-related research has driven the increase in the overall FWCI in China, contributing significantly to China's academic research impact. The FWCI of US nano-publications remained relatively high between 2000 and 2005 and has declined since then. However, the United States still has the highest impact for nano-publications among the key countries. In 2019, China surpassed the United States in the nano-publication FWCI score, but the United States still holds a higher average FWCI than Germany, Japan, the United Kingdom, and the other countries.

The share of citations contributed by nano-publications is increasing

The citation count is a widely adopted direct indicator for measuring the academic impact of scientific research output. It reflects the degree of a scientific publication's impact on other publications. However, the publication's volume can influence the citation count, and the number of citations also increases over time. To reduce this bias, the citation rate (the citation count of the country's nano-publication/overall citation count in the country) was applied to measure nano-publications' citation performance for the world and key countries. Nano-publication citation trends were as follows (Fig. 1.10):

- In key countries and worldwide, the citation rate continued to increase, indicating a growing significance in nano-publications' contribution to the citation system.

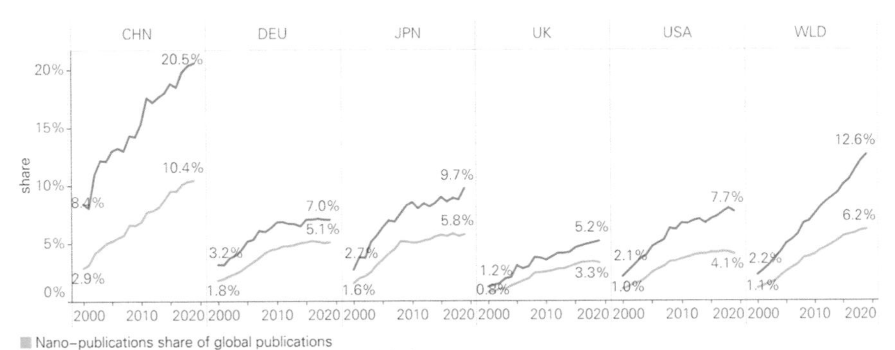

FIGURE 1.10

Percentage of scholarly output and citation count for the world and leading countries (2000–19). *CHN*, China; *DEU*, Germany; *JPN*, Japan; *UK*, United Kingdom; *USA*, United States; *WLD*, world.

From Scopus.

- Nano-publications contributed more to the citation count than to the publication count. For example, whereas 6.2% of publications worldwide were related to nanoscience in 2019, these publications accounted for 12.6% of the global citation count.

About 13.6% of the top 1% highly cited publications worldwide were nano-publications, a much higher percentage than all other fields combined

The top 1% highly cited publications refers to publications with citation counts reaching the top 1% of all publications worldwide on the subject, demonstrating the publishing entity's significant academic impact. In this section, the number and share of nano-publications with citation counts in the top 1% were evaluated to reflect nanoscience's importance in academic research.

Between 2000 and 2019, 11% of the world's top 1% highly cited publications were nano-publications, and the percentage gradually increased every year (Fig. 1.11). In 2000, 4.2% of the top 1% highly cited publications were related to nanoscience.

FIGURE 1.11

Nano-publications' share and trend in top 1% highly cited publications worldwide (2000–19). *CHN*, China; *DEU*, Germany; *JPN*, Japan; *UK*, United Kingdom; *USA*, United States; *WLD*, world.

From Scopus.

For the key comparator countries, nanoscience research has been crucial for the output of high-impact research. In China, 25% of the top 1% most cited publications were related to nanoscience research over the study period, demonstrating the vital contribution of nanoscience to China's academic excellence in research output.

1.4 Analysis of institutional academic research output and impact on nanoscience

Institutions are essential components of nano-research. In this section, we provide an overview and analysis from the perspective of leading institutions with high publication yields.

Top 20 institutions by number of nano-publications

The top 20 institutions[1] with the highest nano-publication output in 2010–2019 were selected for a comparative analysis of their research output and academic impact on nanoscience. Of these top 20 institutions, 11 were from China (Fig. 1.12). Among them, the Chinese Academy of Sciences (CAS) ranked first globally, with 49,441 nano-publications, followed by the French National Center for Scientific Research, with 25,266 nano-publications. The University of California system, including 11 branches, ranked third. The University of the Chinese Academy of Sciences ranked fourth, and the Russian Academy of Sciences fifth.

Academic impact as measured by the FWCI showed that 17 institutions among the top 20 had higher FWCI scores than the global average for nano-publications. The University of Science and Technology of China and Fudan University ranked fourth and fifth, whereas CAS ranked ninth. The positive correlation between the share

[1] Eleven University of California campuses and 98 Max Planck Society institutes were combined to participate in the rankings.

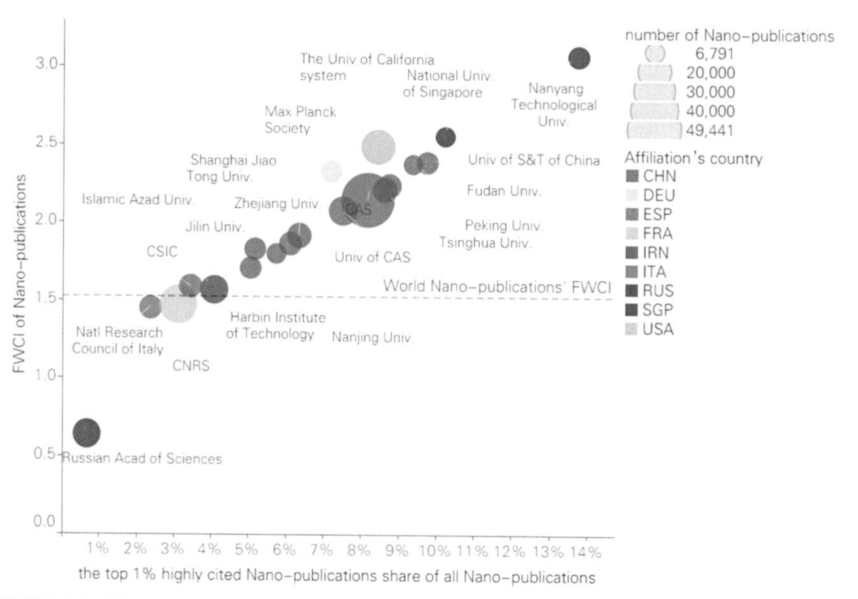

FIGURE 1.12

Scholarly output and academic impact of nano-publications of top 20 institutions (2010–19). *AUS*, Austria; *BRA*, Brazil; *CAN*, Canada; *CHE*, Switzerland; *CHN*, China; *DEU*, Germany; *ESP*, Spain; *FRA*, France; *IND*, India; *IRN*, Iran; *ITA*, Italy; *JPN*, Japan; *KOR*, Korea; *NLD*, The Netherlands; *POL*, Poland; *RUS*, Russia; *SGP*, Singapore; *TWN*, Taiwan; *UK*, United Kingdom; *USA*, United States.

From Scopus.

of the top 1% highly cited publications and the FWCI illustrates the significant role of highly cited publications in enhancing institutions' academic influence.

In addition, research in nanoscience had a crucial role in these institutions' academic impact and research output. Take Chinese institutions as an example (Fig. 1.13), 8% to 17% of their academic research yield was related to nanotechnology, and 13% to 27% of their citations were from nano-publications.

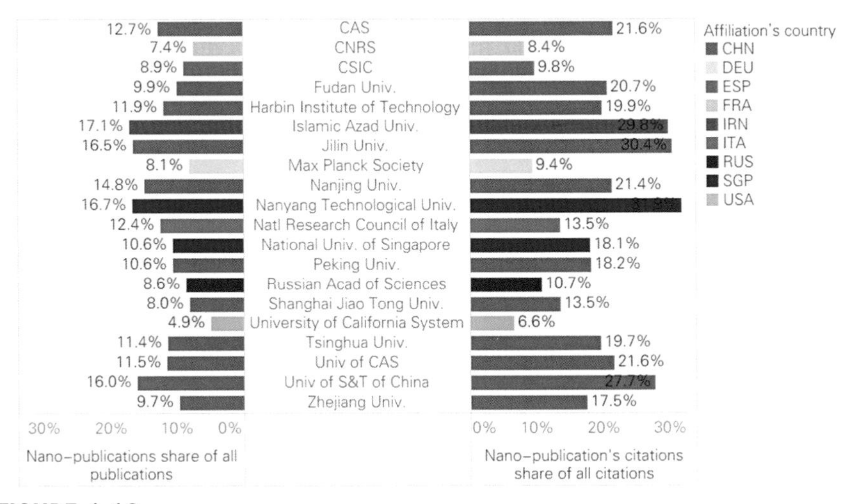

FIGURE 1.13

Nano-publications' share of total scholarly output and citation count at institution level (2010–19), by number of nano-publications. *CAS*, Chinese Academy of Sciences; *CHN*, China; *CNRS*, French National Center for Scientific Research; *CSIC*, Spanish National Research Council; *DEU*, Germany; *ESP*, Spain; *FRA*, France; *IRN*, Iran; *ITA*, Italy; *RUS*, Russia; *SGP*, Singapore; *Univ of S&T*, University of Science and Technology; *USA*, United States.

From Scopus.

Comparison of National Center for Nanoscience and Technology, China and global leading institutions

As well as the top publishing institutions, several of the world's leading nano-research centers were selected for a comparative analysis with the National Center for Nanoscience and Technology (NCNST). Comparison of the academic research output and nano-publication impact between the NCNST and other world-class institutions showed that nano-publications from the NCNST are at the forefront of global academic impact.

In terms of research output, the number of nano-publications from the NCNST ranked in the middle of the group of comparator institutions between 2010 and 2019 (Fig. 1.14), mainly owing to differences in the numbers of authors. Further analysis of research output

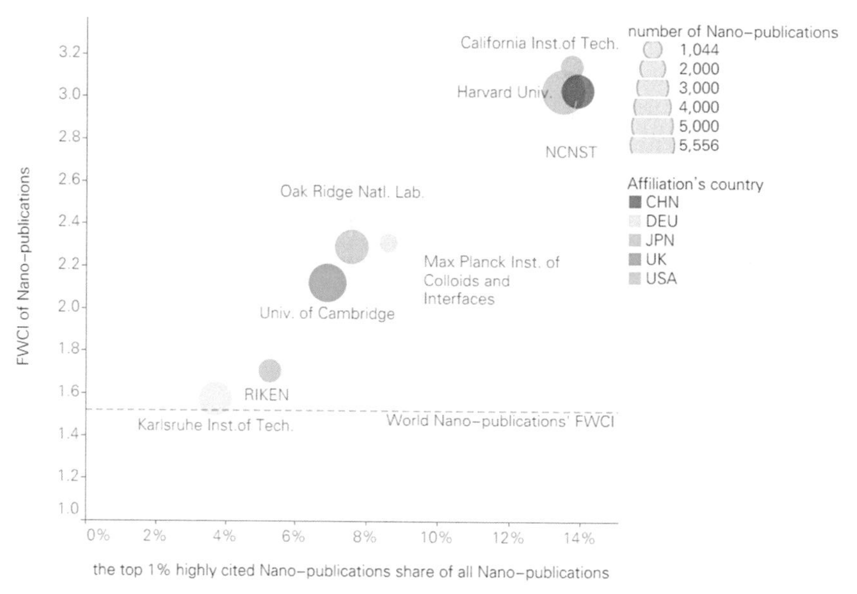

FIGURE 1.14

Comparison of nano-publication research output and academic impacts of the National Center for Nanoscience and Technology and world-class institutions (2010–19). *CHN*, China; *DEU*, Germany; *FWCI*, field-weighted citation impact; *Inst.*, Institute; *JPN*, Japan; *Natl.*, National; *NCNST*, National Center for Nanoscience and Technology; *RIKEN*, Japan Institute of Physical and Chemical Research; UK, United Kingdom; *Univ.*, University; *USA*, United States.

From Scopus.

per capita[2] showed that Oak Ridge National Laboratory, the Max Planck Institute of Colloids and Interfaces, and the NCNST was ranked the top three.

In terms of academic impact, the California Institute of Technology, Harvard University, and the NCNST ranked top three for nano-publication FWCI (Fig. 1.14, *y*-axis) between 2010 and 2019. The NCNST held the highest percentage in the top 1% highly cited nano-publications (Fig. 1.14, *x*-axis) among the comparator institutions. The analysis showed that nano-related academic outputs from the NCNST, the California Institute of Technology, and Harvard University had a relatively higher academic impact.

[2] Per capita = total output from the institution/count of authors from the institution.

CHAPTER

2 The contribution of nanoscience to basic research

Key findings

Physical sciences
In the subject area of physical sciences, nano-publications accounted for the largest percentage among all scholarly outputs between 2000 and 2019. Of all publications, over 12% of those in the subjects of materials science, chemistry, physics and astronomy, and chemical engineering were related to nanoscience. Materials science was the subject with the most nano-publications.

Engineering, energy, and pharmacology, toxicology, &

Life sciences
Nano-publications had the fastest growth in the subject area of life sciences, although the volume was small. Between 2000 and 2019, the annual growth rate of nano-publications was 5.1 times the average in the subjects of immunology and microbiology; 4.4 times the average in biochemistry, genetics, and molecular biology; and four times the average in pharmacology, toxicology, and pharmaceutics.

Subject differences between China and the United States
In 2000–19, China

Big Data Analysis of Nanoscience Bibliometrics, Patent, and Funding Data (2000–2019)
https://doi.org/10.1016/B978-0-323-91311-9.00002-5

pharmaceutics
Nanoscience significantly boosted the academic research impact of multiple subject areas, especially in engineering, energy, and pharmacology, toxicology, & pharmaceutics. The field-weighted citation impact of nano-publications between 2000 and 2019 in engineering was 1.9 times the average in the subject of engineering; it was 1.8 times the average in pharmacology, toxicology, and pharmaceutics, and 1.75 times the average in energy.

had a larger nano-publication share than the United States and the world in the subjects of physics and astronomy, chemistry, chemical engineering, and pharmacology, toxicology, and pharmaceutics. The United States had a larger nano-publication share than China and the world in energy, engineering, and materials science.

89%
In 2015–19, 89% of the world's most prominent research topics had at least one nano-publication in the topic's publication set.

39%
In 2015–19, 39% of the world's most prominent topics were strongly related to nanoscience, with a share of 10% or more publications in nano-research.

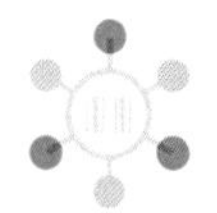

Top output prominent topics for nanoscience
In 2015–19, nanoscience recorded the highest academic research

Fast growing prominent topics for nanoscience
In 2015–19, nano-publications showed rapid growth in the most prominent topic

output in the most prominent topic clusters,[1] including solar cells, graphene, lithium battery, plasma metamaterials, biosensor, catalysts, and semiconductor quantum dots.

clusters,[2] including DNA sequencing and tumor treatment, wastewater treatment, cellulose, organic metal, activated carbon, water purification/desalination, and quantum computing.

In this chapter, nanoscience's universality in basic science is illustrated by knowledge flow diagrams, and nanoscience's contribution to each field is evaluated by academic research output and impact. The chapter provides an analysis of nanoscience's implications for basic science research in each key country. The scientific disciplines in this chapter are categorized according to the Scopus All Science Journal Classification (ASJC) codes, in which scientific activity is classified into four subject areas, 27 subjects, and 334 fields. See Appendix D for a breakdown of the classification. The analysis is focused on eight key subjects with the highest share of nano-publications in the subject, and one subject (immunology and microbiology) as a highly relevant subject for nanoscience.

2.1 Universality: nanoscience in basic science

Knowledge flow diagram of nanoscience

Based on the subject distribution of referenced and cited publications between 2015 and 2019, a knowledge flow diagram[3] of nanoscience is

[1] The topic cluster's prominence score was in the top 5% globally and had the most nano-publications.

[2] The topic cluster's prominence score was in the top 5% globally and had the highest CAGR of nano-publications. To exclude topics with a small volume but extremely high growth rate, here we counted only topics with at least 2000 nano-publications in 2015–19. To identify emerging areas, the top 10 output topics for nanoscience were excluded, which were already mature areas in nanoscience.

[3] Some ASJC subjects were merged in the diagram; more details are listed in the Appendix.

presented (Fig. 2.1). The diagram provides a macroscopic illustration of nanoscience's impact on the basic scientific subjects. In general, nanoscience was widely distributed in the subject area of physical sciences and gradually penetrated into the subject areas of life sciences and health sciences, indicating nanoscience's emergence into multiple fields as a universal science.

The sources of knowledge of most nano-publications, as shown by their references, were the subjects of chemical engineering and chemistry; materials science; physics and astronomy; engineering; biochemistry, genetics, and molecular biology; and energy. The disciplinary distribution of the references was similar to that of the nano-publications themselves. In addition, with the development of nanoscience in life sciences and health sciences, the scope of nano-publication references also extended into various subjects of health sciences; environmental science; and pharmacology, toxicology, and pharmaceutics, as well as other fundamental sciences.

Knowledge in nano-publications flowed out to the subjects of chemical engineering and chemistry; materials science; engineering;

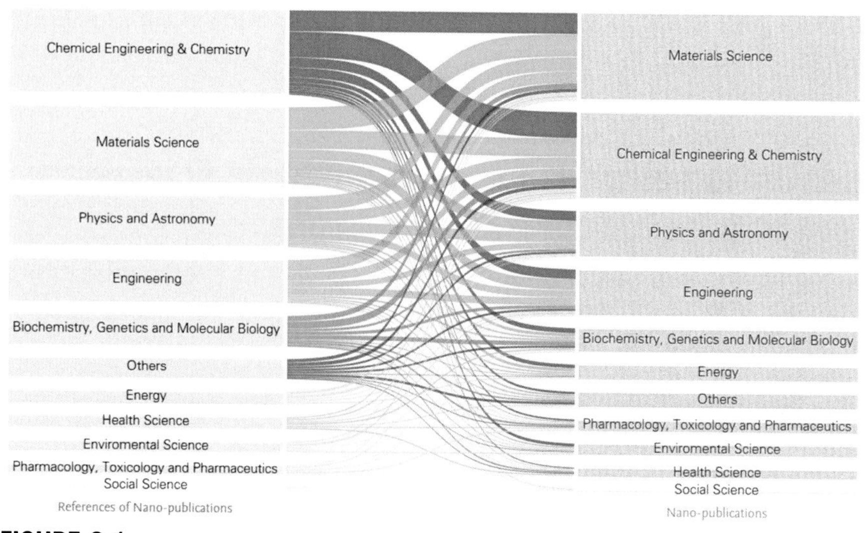

FIGURE 2.1

Knowledge flow-in diagram of nano-publications between All Science Journal Classification subjects (2015–19).

Source: Scopus.

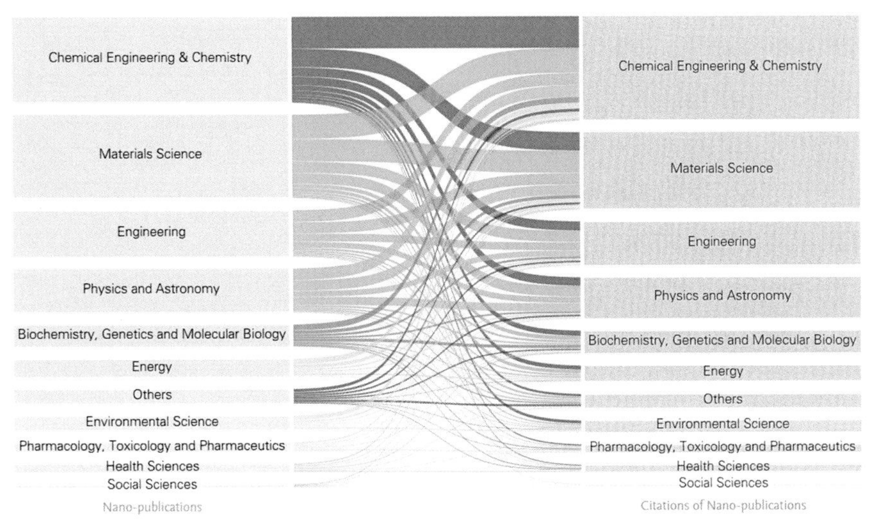

FIGURE 2.2

Knowledge flow-out diagram of nano-publications between All Science Journal Classification subjects (2015–19).

Source: Scopus.

physics and astronomy; biochemistry, genetics, and molecular biology; and energy (Fig. 2.2). Nano-publications were also cited by publications in the subjects of environmental science; pharmacology, toxicology, and pharmaceutics; and subjects in health sciences, such as medicine and dentistry, although in smaller proportions.

Nanoscience's contribution to scholarly output of basic science

The volume, share, and compound annual growth rate (CAGR) of nano-publications illustrate the impact of nanoscience on various disciplines. Nanoscience was widely applied in the subject area of physical science between 2000 and 2019 (Fig. 2.3). Within this subject area, the subjects of materials science, chemistry, physics and astronomy, engineering, and chemical engineering had the most nano-publications. Focusing on the share of nano-publications can minimize differences in scale among varying fields. However, in physical science, nano-publications still made up a high percentage of literature. Nano-publications were attributed to 20.7%, 17.7%, and 16.3% of

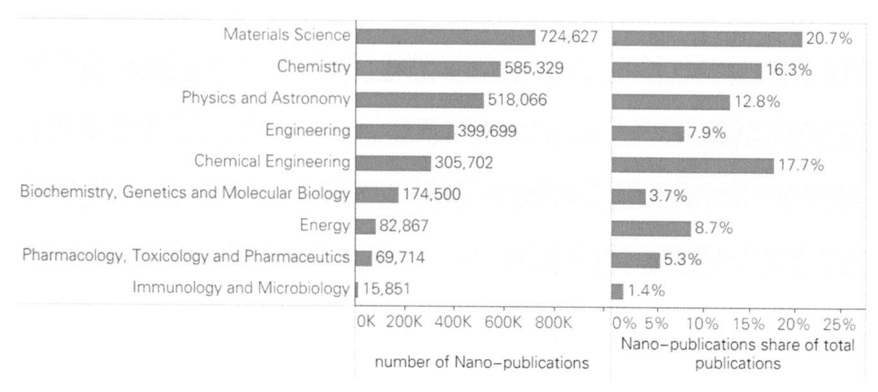

FIGURE 2.3

Nano-publications' share of scholarly output in key subjects (2000–19).

Source: Scopus.

total articles in materials science, chemical engineering, and chemistry, respectively.

The CAGR of nano-publications was higher than average in each comparator subject (Fig. 2.4, with a relative CAGR of >1[4]) in 2000–19. Among them, the growth rate of nano-publications in immunology and microbiology was 5.1 times the subject's average.

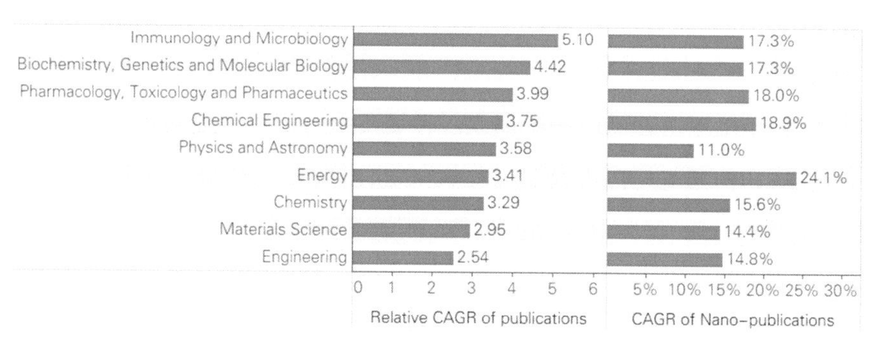

FIGURE 2.4

Relative compound annual growth rate (CAGR) and CAGR of nano-publications in key subjects (2000–19).

Source: Scopus.

[4] Relative CAGR = CAGR of nano-publications in the subject/CAGR of all publications in the subject.

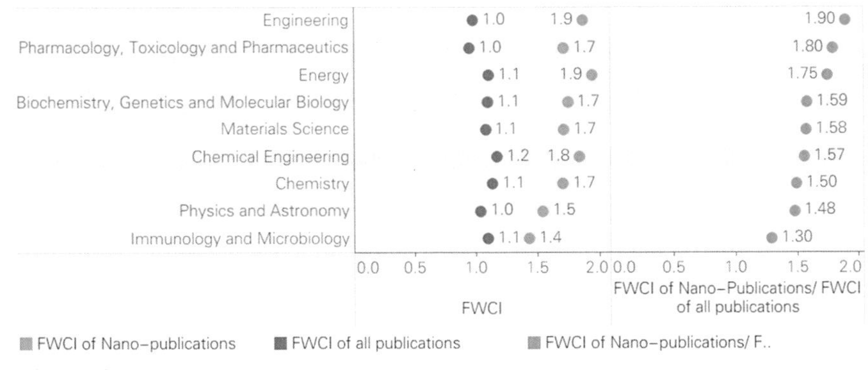

FIGURE 2.5

Comparison of nano-publications' field-weighted citation impact (FWCI) and average FWCI in key subjects (2000–19).

Source: Scopus.

Nanoscience's contribution to academic impact of basic science

The field-weighted citation impact (FWCI) of nano-publications was higher than the FWCI for all publications in each analyzed subject (Fig. 2.5). Among them, the FWCI of engineering increased the most. Between 2000 and 2019, the FWCI for nano-publications in engineering was 1.9 times the average, followed by 1.8 times the average in pharmacology, toxicology, and pharmaceutics, and 1.75 times in energy.

Nanoscience's contribution to basic science in key countries

Scholarly output

Nano-publication shares in materials science, chemical engineering, and chemistry were the highest in the world between 2000 and 2019 (Fig. 2.6). The United States had a higher nano-publication share than China and the world in energy, engineering, and materials science. China held a higher percentage of nano-publications than the United States and the rest of the world in physics and astronomy, chemistry, chemical engineering, and pharmacology, toxicology, and pharmaceutics. Immunology and microbiology had the smallest share of publications in nanoscience of the subjects, but its relative growth rate was

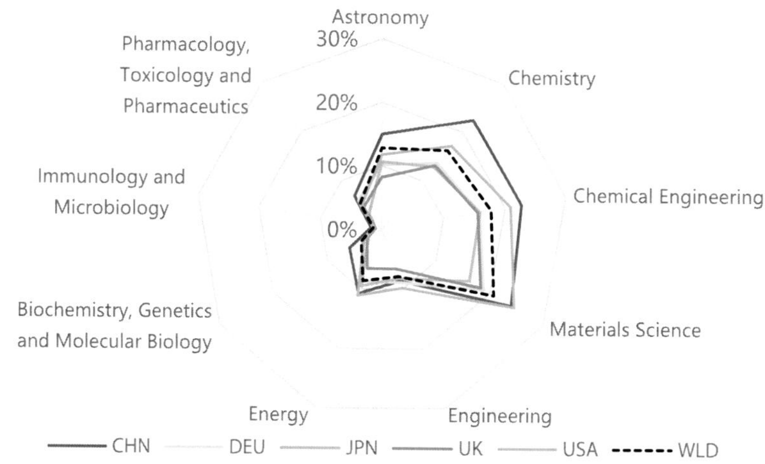

FIGURE 2.6

Nano-publications' share in key subjects in each comparator country and the world (2000–19). *CHN*, China; *DEU*, Germany; *JPN*, Japan; *UK*, United Kingdom; *USA*, United States; *WLD*, world.

Source: Scopus.

higher than those of the other subjects (Fig. 2.4), indicating the trend of disciplinary integration between nanoscience and this subject.

Fig. 2.7 illustrates the share of nano-publications in key subjects, for each key country and the world. It shows that the percentage of nano-publications was growing in all nine subjects between 2000 and 2019 for all key countries and the world. However, the growth rates of these publications varied slightly across subjects and countries.

Academic impact

The FWCI of nano-publications in each subject listed subsequently was higher than the respective subject's overall average FWCI, indicating that nanoscience has enhanced the academic impact in multiple subjects in each key country (Fig. 2.8).

Especially in China, nanoscience boosted the academic impact of various subjects. Between 2000 and 2019, in most of the subjects subsequently, the FWCI of China was lower than those of the United States, the European countries, and the world average. However, as nanoscience became integrated into these subjects, China's academic impact started to rise. The increase in FWCI was especially significant in the energy and biochemistry, genetics, and molecular biology subjects.

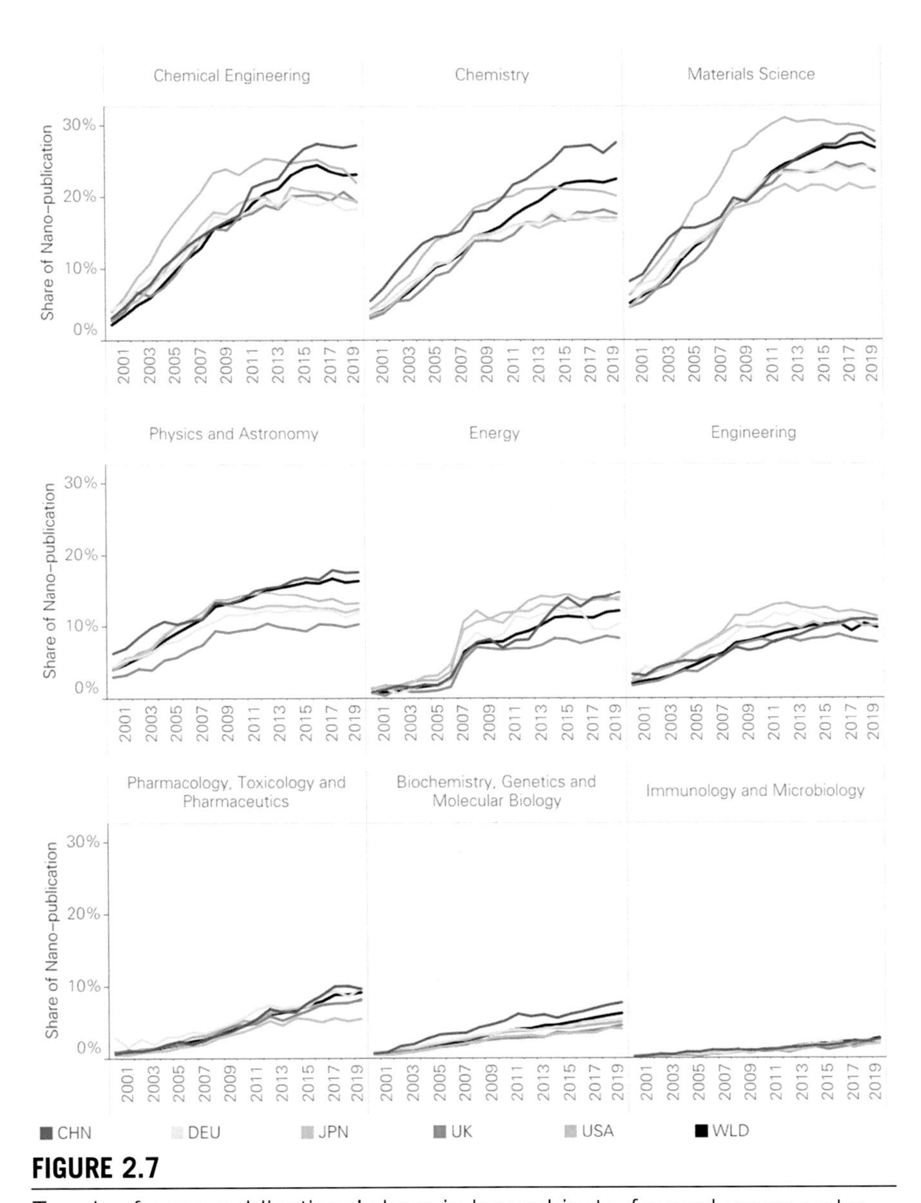

FIGURE 2.7

Trends of nano-publications' share in key subjects, for each comparator country and the world (2000–19). *CHN*, China; *DEU*, Germany; *JPN*, Japan; *UK*, United Kingdom; *USA*, United States; *WLD*, world.

Source: Scopus.

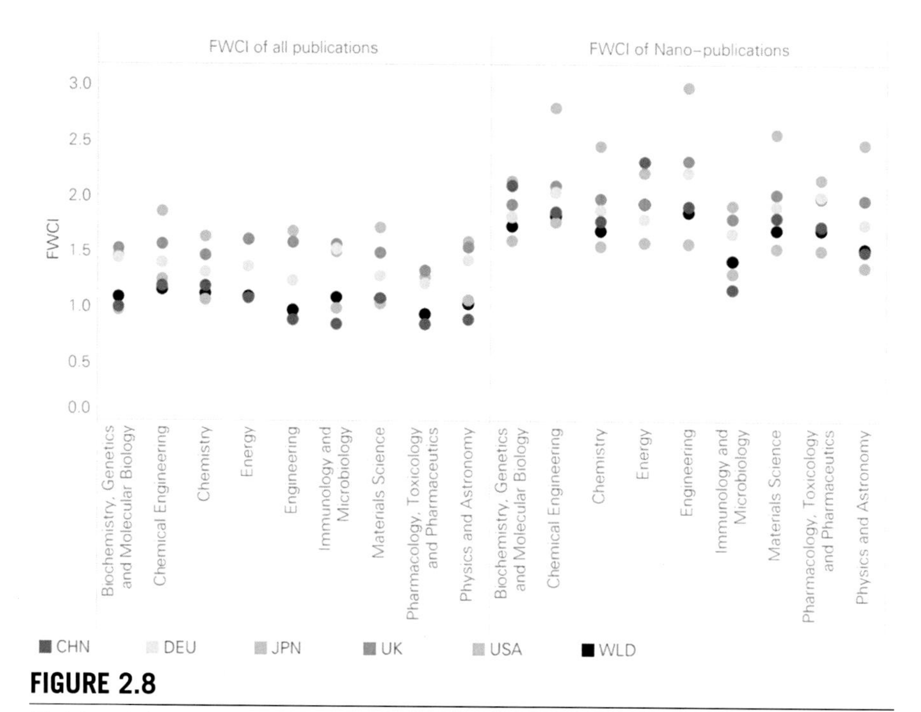

FIGURE 2.8

Field-weighted citation impact (FWCI) of overall publications and nano-publications in key subjects in each country (2000–19). *CHN*, China; *DEU*, Germany; *JPN*, Japan; *UK*, United Kingdom; *USA*, United States; *WLD*, world.

Source: Scopus.

2.2 Advanced science: nanoscience in highly prominent topics

The term “topics” refers to nearly 96,000 research topics created using the citation patterns of Scopus-indexed publications. A topic is a dynamic collection of documents with a common focused intellectual interest. The methodology for using citation patterns to define research topics was developed through an Elsevier collaboration with research partners. A topic is created in which linkages within the topic are strong and linkages outside the topic are weak.[5] This methodology

[5] Learn about topics and topic clusters: Topic Prominence in Science FAQs - SciVal Support Center.

considers 95% of articles available in Scopus from 1996 to the present and clusters them into nearly 96,000 unique, global research topics based on citation patterns.

The Prominence score is an indication of the momentum related to a particular topic. Prominence is a linear combination of citations, views, and journal impact for a given topic, in which each factor is normalized by the topic's standard deviation. More details can be found in Appendix B. The Prominence score has been shown to correlate with topic-level funding per author in a US sample of funded grants. The analysis found that on average, the higher the Prominence score, the more money per US author was available for research on that topic.[6]

Topic clusters are a higher-level aggregation of these research topics based on the same direct citation algorithm that creates the topics. Although topics are easy to understand for their respective subject experts, they are more difficult for subject generalists. To aid discovery and understanding of the topics, we have taken the topics and aggregated them into around 1500 topic clusters. When the strength of the citation links between topics reaches a threshold, a topic cluster is formed.

In this section, we will analyze the performance of nanoscience in the world's most prominent topics.

Nanoscience's connection with the most prominent topics

The world's top 1% most prominent topics represent research areas with the most attention and popularity. In total, there are around 960 topics with the Prominence score ranked in the top 1% worldwide.

Based on the percentage of nano-publications in highly prominent topics, nanoscience's contribution and correlation to these topics were evaluated in this study. The number of topics and their different proportions of nano-publications are aggregated in Table 2.1:

[6] Richard Klavans, Kevin W. Boyack, Research portfolio analysis and topic prominence, Journal of Informetrics, Volume 11, Issue 4,2017, pages 1158–1174, ISSN 1751–1577, https://doi.org/10.1016/j.joi.2017.10.002.

Table 2.1 Distribution of nano-publications shares in top 1% high-prominence topics (2015–19).

Proportion of nano-publications in most prominent topics	Number of world's top 1% high-prominence topics	Share of all top 1% high-prominence topics (2015–19)
Nano-publications share in topic: ≥90%	26	3%
Nano-publications share in topic: 70–90%	68	7%
Nano-publications share in topic: 50–70%	63	7%
Nano-publications share in topic: 30–50%	67	7%
Nano-publications share in topic: 10–30%	151	16%
Nano-publications share in topic: 5–10%	68	7%
Nano-publications share in topic: 1–5%	188	20%
Nano-publications share in topic: <1% (>0%)	226	24%
No nano-publications in topic	102	11%

From Scopus, SciVal.

- Among the world's most prominent topics globally, 89% were related to nanoscience, meaning they had at least one nano-publication. These topics were widely distributed across multiple subject areas (Fig. 2.9).
- Up to 40% of the world's most prominent topics strongly correlated to nanoscience, meaning the share of nano-publications in these topics was 10% or more. Among them, 26 topics had a nano-publication percentage of over 90%.

By analyzing the correlation between nanoscience and each subject area's hottest topics, which are prominent topics that rank in the top 1% globally, we can better understand nanoscience's impact on these topics.

Based on our analysis, in 18 of the 27 ASJC subjects,[7] over 70% of the most prominent topics had at least one nano-publication. Furthermore, 16 of the ASJC subjects strongly related to nanoscience, meaning that the topics had 10% or more nano-publications. As shown in Fig. 2.10, nanoscience was strongly correlated to the most

[7] Scopus has 27 ASJC subject areas.

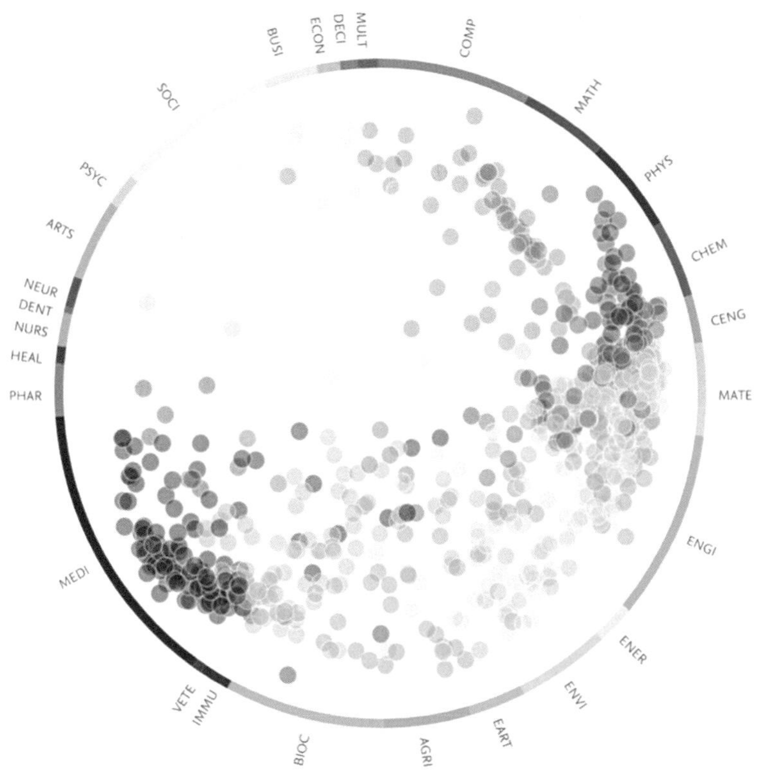

FIGURE 2.9

Subject distribution of the world's top 1% high prominence topics with nano-publications (2015–19). Each *dot* represents a topic; a *dot's color* represents the subject area to which the dot belongs.

Source: Scopus, SciVal.

prominent topics[8] in materials science (85.4% of the most prominent topics in materials science have at least 10% of publications related to nanoscience), physics and astronomy (78.9%, respectively), chemistry (76.7%, respectively), chemical engineering (73.2%, respectively), engineering (64.2%, respectively), energy (50.0%, respectively), and pharmacology, toxicology, and pharmaceutics (42.9%, respectively).

[8] Topic prominence score ranking in the top 1% worldwide.

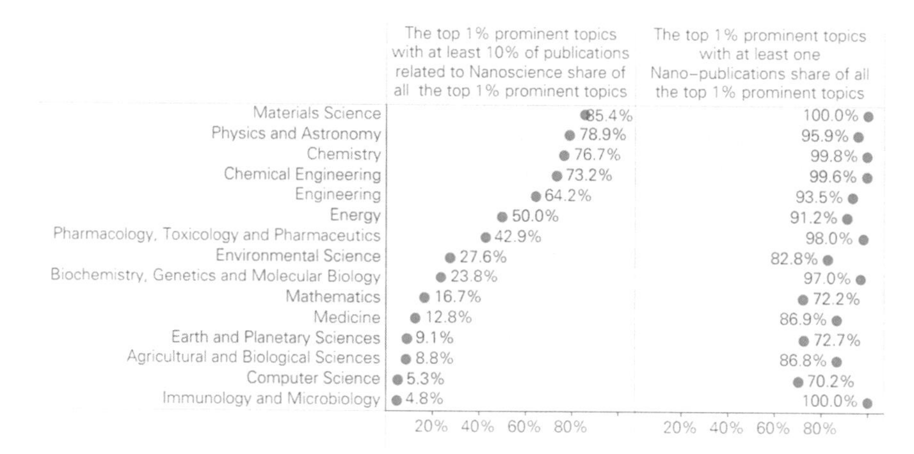

FIGURE 2.10

Correlation analysis of nanoscience and high-prominence topics in each subject area in the world (2015–19).

Source: Scopus.

The most prominent topic clusters related to nanoscience

Topic clusters are aggregations of topics with similar interests that form a broader, higher-level area of research. These topic clusters can provide a deeper insight into ongoing research done by countries, institutions, and scientists, before drilling down into the underlying niche topics. In this section, we analyzed prominent nano-related topic clusters, which were those ranked in the top 5% in the world, based on the scale and growth of their scholarly output.

The highly prominent topic clusters with the most nano-publications (2015–19)

Between 2015 and 2019, the top 5% of worldwide topic clusters by prominence[9] with the most nano-publications were solar cells, graphene, lithium batteries, plasmonic metamaterial, biosensors, catalysts,

[9] Topic clusters with prominence scores ranking in the top 5% globally.

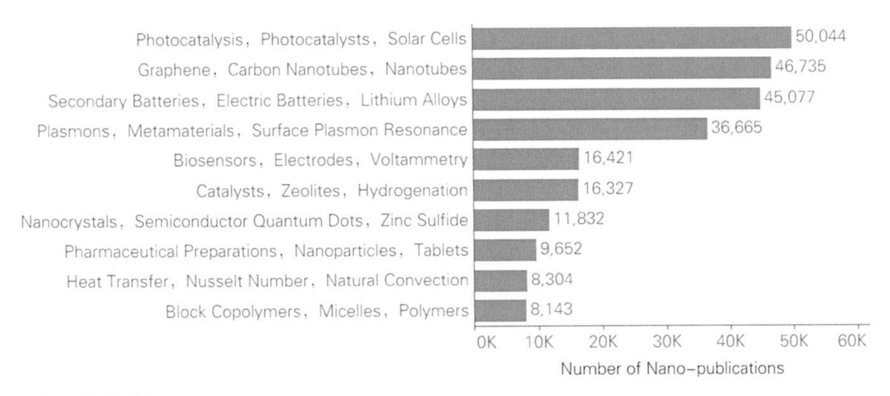

FIGURE 2.11

Top 10 topic clusters with a prominence score in the top 5%, by nano-publications in topic clusters (2015–19).

Source: Scopus, SciVal.

semiconductor quantum dots, pharmaceutical nanoparticles, and polymers (Fig. 2.11).

In the topic cluster "Photocatalysis, Photocatalysts, Solar Cells," the term "nano" chiefly relates to nanorods, perovskite, and solar cells. The five publications with the highest citations in this topic cluster were all related to perovskite solar cells.

Nanoscience application in the "Biosensors, Electrodes, Voltammetry" topic cluster was mainly reflected in the combination of biosensor, electrochemistry and graphene, nanoparticles, glassy carbon, and similar topics. Studies representing the cluster included electrochemical sensors and biosensors based on nanomaterials and nanostructures, electrochemistry of graphene-related materials, and so forth.

Applications of nanoscience in the "Pharmaceutical, Preparations, Nanoparticles, Tablets" topic cluster were mainly related to nanotechnology in the pharmaceutical industry. Research that represented the cluster included nanoparticles applications and the nanostructured lipid carrier in drug delivery systems.

The Highly prominent topic clusters with the highest growth rate, by nano-publication output (2015–19)

The 10 highly prominent topic clusters[10] (Fig. 2.12) with the fastest-growing nano-publication output demonstrated the dynamic application of nanoscience in DNA sequencing and tumor treatment, wastewater treatment, cellulose, metal-organic framework, activated carbon, water purification/desalination, quantum computing, and so forth. The exponential growth in publication count has reflected the integration and rapid development of nanoscience in emerging research areas.

Among the top 10 topic clusters, "MicroRNAs, Long Untranslated RNA, Neoplasms" had the fastest-growing nano-publication output. Studies that represent the cluster include applying nanotechnologies in the fields of DNA sequencing, peptide detection, microRNA detection, and delivery carriers for targeted therapeutic drugs.

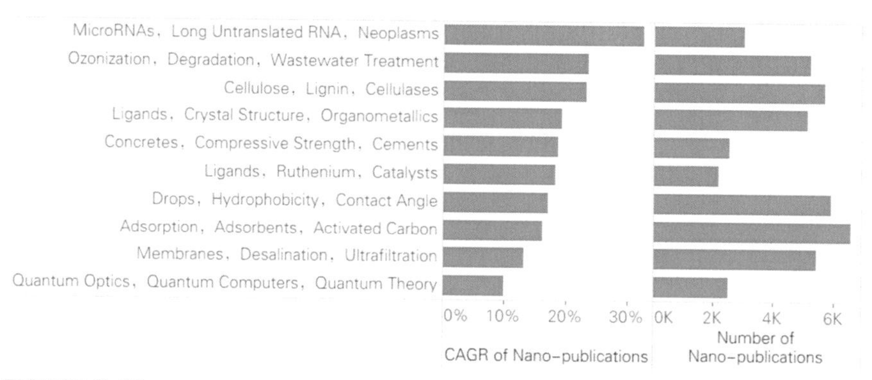

FIGURE 2.12

Top 10 topic clusters with a prominence score in the top 5%, by compound annual growth rate (CAGR) of nano-publications in the topic clusters (2015–19).

Source: Scopus, SciVal.

[10] The topic cluster's prominence score was in the top 5% globally and had the highest CAGR of nano-publications. To exclude topics with a small volume but extremely high growth rate, here we counted only topics with at least 2000 nano-publications in 2015–19. To identify emerging areas, the top 10 output topics for nanoscience were excluded, which were already mature areas in nanoscience.

In addition, there were three topic clusters related to water treatment: ozonization, degradation, and wastewater treatment. They included studies such as "Environmental Remediation and Application of Nanoscale Zero-Valent Iron," and "Drops, Hydrophobicity, Contact Angle." There were representative research in nanomaterials and nanostructure, such as the hydrophilicity and hydrophobicity of nanobubbles and nanodroplets; and membranes, desalination, and ultrafiltration. Studies investigating the applications of nanofiltration, nanofilms, and nanofiltration membranes in water desalination and purification were also included. These hot research areas reflect the increasing development of nanoscience in environmental management, especially in water treatment.

CHAPTER

The impact of nanoscience on industry

3

Key findings

2.6%
of worldwide nano-publications involved academic and corporate collaborations in 2015–19. The number was 0.2% lower than the global average of 2.8% for all publications.

1.5%
of Chinese nano-publications involved academic and corporate collaborations in 2015–19, which was lower than the overall average in China. In the United Kingdom, the United States, Japan, and Germany, academic–corporate collaboration rates in nanoscience were higher than the overall average in each country.

CNRS, Samsung, Chinese Academy of Sciences (CAS), IBM, and Sinopec
were the top five institutions with the highest number of academic–corporate collaborated nano-publications between 2015 and 2019.

1.04%
of nano-publications were cited at least once by patents in the top five international patent libraries in the world, which was 89% ahead of the all-fields average (2015–19).

Big Data Analysis of Nanoscience Bibliometrics, Patent, and Funding Data (2000–2019)
https://doi.org/10.1016/B978-0-323-91311-9.00003-7

693,000
patents[1] were related to nanoscience in 2000–19, accounting for 2% of patents worldwide. Among them, 58% came from China.

CAS, Tsinghua U, and MIT
were the top three academic institutions with the highest nano-related Patent Asset Index (2000–19).

Samsung, LG Chem, and Foxconn
were the top three corporate entities with the highest nano-related Patent Asset Index (2000–19).

Based on Scopus data, 17,326 nano-publications were published worldwide via academic–corporate collaborations between 2015 and 2019, accounting for 2.6% of all nano-publications. The rapid growth of China's overall nano-publications drove the development of academic–corporate collaborative nano-research in the nation. However, in contrast to other key countries, China's share of academic–corporate collaborative nano-publications among all of its nano-publications was still relatively low.

The 100 global corporate entities with the most nano-publications produced in collaboration with academic research institutions were located in the countries and regions with the most advanced nanoscience research. These corporations belonged to a range of industries that varied among countries, indicating diversified nanoscience application across industries. The corporate entities involved in nanoscience research in China were mainly petrochemical companies, whereas those in the United States were mostly high-tech companies such as IBM, Intel, and Thermo Fisher.

Research in basic science was the driving force for science and technology innovations and also the source for applied science and

[1] Nano-related patents: obtained by searching "Nano*" in the title, abstract or claim of patents. Data source: PatentSight. Search time: Mar. 30, 2020. Filing year: 2000–19.

breakthroughs. The growing share of nanoscience research output has shifted from basic research to industrial application, in which the percentage of patent-cited publications was higher than in other fields.

Patents can foretell the application prospects of a certain technology in the industry. Between 2000 and 2019, a total of 693,789 nano-related patents were granted worldwide, 58% of which were from China and 12% from the United States. The rapid growth of China's nano-related patents has driven the surge of nano-related patents globally. However, compared with the United States and European countries, the competitiveness of nano-related patents from China still has space for improvement.

3.1 Analysis of academic–corporate collaborations in nanoscience

Academic–corporate collaborative research output in nanoscience

Collaborations between academic institutions, mainly universities and research centers, and corporate research and development (R&D) centers often contributes to the basic research knowledge flow between academia and industry. The flow increases the opportunity for technology transfer and helps to secure funding for academia. In this section, academic–corporate collaborative research outputs between 2015 and 2019 are analyzed to evaluate cross-sectoral nano-research between academic institutions and corporate entities.

Number of academic–corporate collaborative nano-publications

Based on Scopus data, a total of 17,326 nano-publications were published with academic–corporate collaborative between 2015 and 2019, accounting for 2.6% of all nano-publications in that period. During that time, the academic–corporate collaboration rate across all research fields was 2.8%. With 6,124 publications, the United States contributed 35% of academic–corporate nano-publications, the most of all countries. The number of Chinese academic–corporate nano-publications increased rapidly. Between 2015 and 2019, the compound annual growth rate (CAGR) of Chinese academic–corporate nano-publications was 14.2%, the highest rate of all comparator countries (Fig. 3.1).

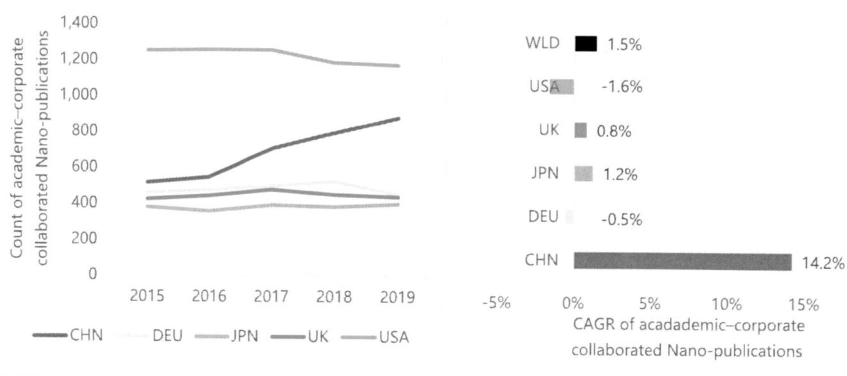

FIGURE 3.1

Trends in research output and compound annual growth rate (CAGR) of academic–corporate collaborated nano-publications (2015–19). *CHN*, China; *DEU*, Germany; *JPN*, Japan; *UK*, United Kingdom; *USA*, United States; *WLD*, world.

Source: Scopus.

Academic–corporate collaboration rate

The increasing number of academic–corporate collaborative nano-publications was tightly connected to overall development in nano-science. The academic–corporate collaboration rate, which is the share of academic–corporate collaborative publications in all publications, can be used to evaluate the progress of cooperation efforts in nanoscience.

The global academic–corporate collaboration rate in nanoscience was lower than the overall average rate for the world (Fig. 3.2). However, in the United States, Japan, Germany, and the United Kingdom, each country's academic–corporate collaboration rate was higher than its overall average, indicating the close relationship between academia and industry in the nanoscience research. China's academic–corporate collaboration rate in nano-research was lower than its national rate as well as the rates of comparator countries and the global average. The academic–corporate collaboration rate for Chinese nano-publications also barely increased between 2015 and 2019.

In terms of research fields, generally, academic–corporate collaboration rates in the areas of medicine and life sciences, such as pharmacology, toxicology and pharmaceutics; immunology and microbiology;

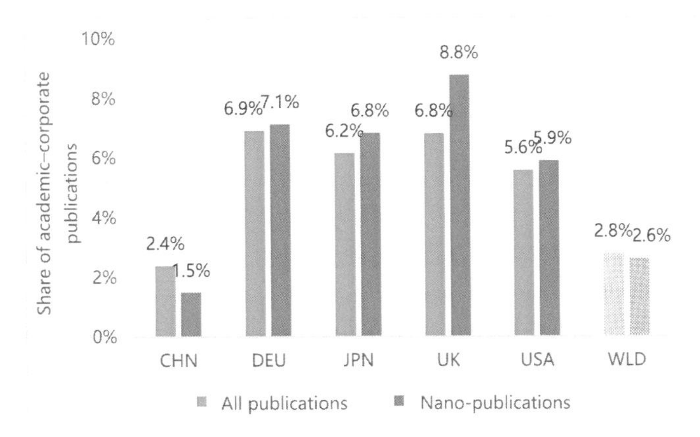

FIGURE 3.2

Comparison of nano-publications' academic–corporate collaboration rate and overall average for each country and the world (2015–19). *CHN*, China; *DEU*, Germany; *JPN*, Japan; *UK*, United Kingdom; *USA*, United States; *WLD*, world.

Source: Scopus.

and biochemistry, genetics, and molecular biology, were higher than those in other subject areas (Fig. 3.3). For example, the United Kingdom had a 13.7% share in academic–corporate nano-publications in pharmacology, toxicology, and pharmaceutics.

The degree of academic–corporate collaborations varied across fields. To avoid disparities caused by disciplinary differences, the book evaluated the degree of nano-research collaboration efforts in each country and the world by the relative academic–corporate collaboration rate[2]. The assessment revealed that Japan had a significantly higher nano-publication academic–corporate collaboration rate in biochemistry, genetics, and molecular biology; immunology and microbiology; and medicine than the average academic–corporate collaboration rate in these subjects (Fig. 3.4).

[2] Relative academic–corporate collaboration rate = academic–corporate collaboration rate of nano-publications in the subject/academic–corporate collaboration rate of the subject. If the value is equal to 1, it means the degree of academic–corporate collaboration in nanoscience is same as the average degree of all research fields.

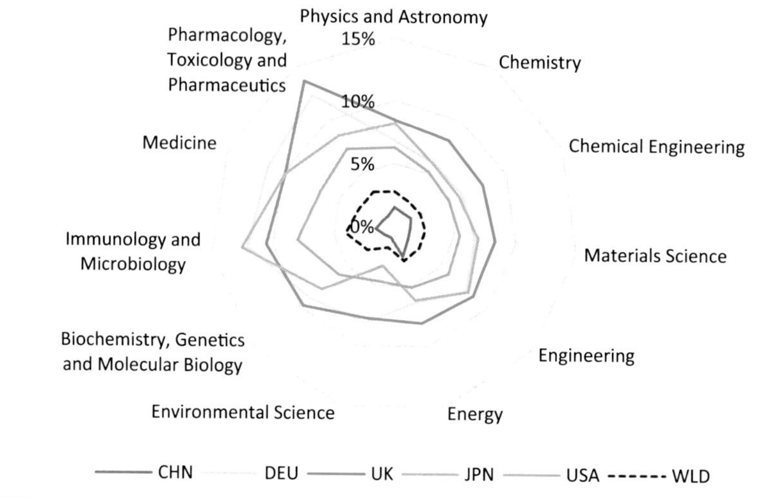

FIGURE 3.3

Academic–corporate collaboration rates of nano-publications in key subjects for comparator countries and the world (2015–19). *CHN*, China; *DEU*, Germany; *JPN*, Japan; *UK*, United Kingdom; *USA*, United States; *WLD*, world.

Source: Scopus.

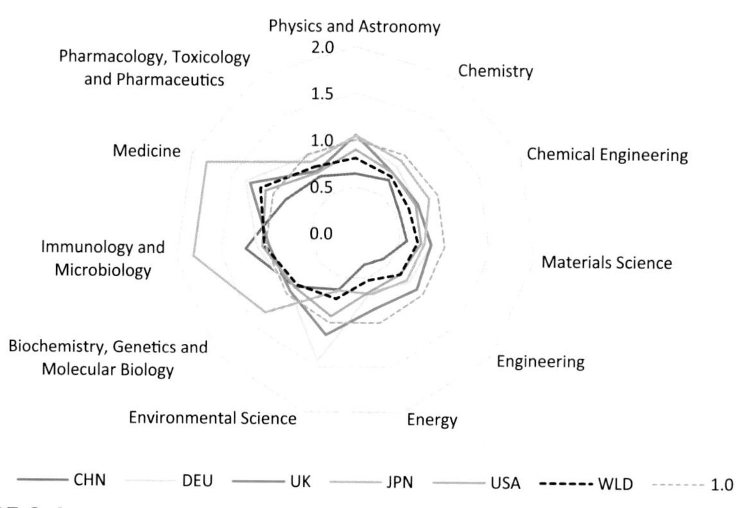

FIGURE 3.4

Relative academic–corporate collaboration rates of nano-publications in key subjects for the comparator countries and the world (2015–19). *CHN*, China; *DEU*, Germany; *JPN*, Japan; *UK*, United Kingdom; *USA*, United States; *WLD*, world.

Source: Scopus.

Academic–corporate collaboration network based on coauthorship

Fig. 3.5 illustrates the world's top 100 corporate entities with the most nano-publications produced in collaboration with academic institutions. The statistics show that these corporations were concentrated in countries and regions with relatively advanced nanoscience research. Among them, corporate entities in the United States were the largest contingent (27 entities in the top 100 corporate list), followed by China (15), Germany (9), Japan (7), Switzerland (5), the Netherlands (5), France (5), and South Korea (4).

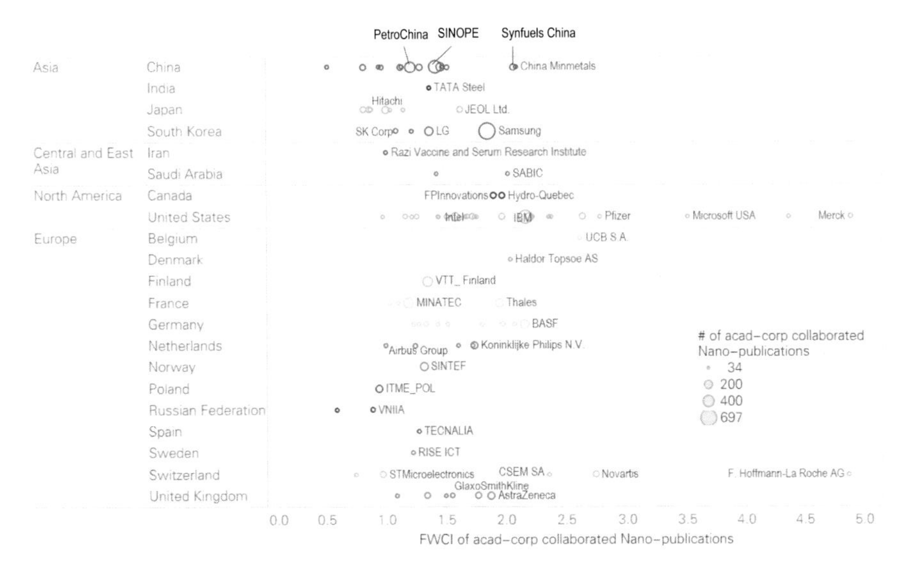

FIGURE 3.5

Top 100 corporate entities by academic–corporate (acad-corp) collaborated nano-publications (2015–19). (Only the some entities' names are marked; the full list is presented in Appendix A). *FWCI*, field-weighted citation impact.

Source: Scopus.

Samsung, IBM, and Sinopec were the top three corporate entities with the most nano-publications copublished with academic institutions (Fig. 3.6). Furthermore, corporations that partnered with academic institutions for nanoscience research came from diverse industries in different countries. For example, most of these top corporate entities were pharmaceutical companies in the United Kingdom such as GSK AstraZeneca; petrochemical companies in China such as Sinopec, CNPC, and RIPED; high-tech companies in the United States such as IBM, Intel, and Thermo Fisher; pharmaceutical and chemical engineering enterprises in Germany such as BASF; electronics enterprises in South Korea such as Samsung and LG; manufacturing companies in Japan such as Hitachi, Toyota Motor Corporation, and JEOL; and semiconductor and pharmaceutical companies in Switzerland such as Novartis and Roche Pharma.

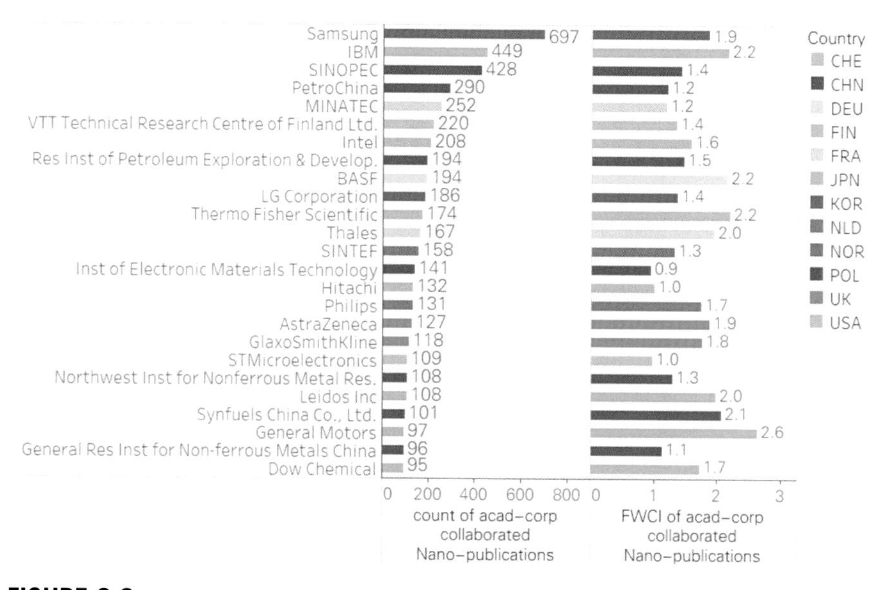

FIGURE 3.6

Top 25 corporate entities[3] with the most academic–*corporate* (acad-corp) collaborated nano-publications (2015–19). *CHE*, Switzerland; *CHN*, China; *DEU*, Germany; *FIN*, Finland; *FRA*, France; *FWCI*, field-weighted citation impact; *JPN*, Japan; *KOR*, Korea; *NLD*, The Netherlands; *NOR*, Norway; *POL*, Poland; *UK*, United Kingdom; *USA*, United States.

Source: Scopus.

In terms of academic impact, as measured by the field-weighted citation impact (FWCI) of nano-publications coauthored by industry and academia, corporate entities from the United States were ahead of those from other countries. China's corporate entities had a relatively weaker

---

[3] Thermo Fisher Scientific Inc. is a US-based biotechnology and medical instruments company. It was formed by the merger of two US biotechnology companies on May 14, 2006 and is headquartered in Waltham, Massachusetts. It mainly produces laboratory equipment, reagents, analytical instruments, consumables, and software. Thales Group is a French electronics group focusing on aerospace, defense, ground transportation, security, and manufacturing electrical systems. The company is headquartered in France; its R&D arms are located in Silicon Valley, France, Paris, and Russia. Since the acquisition of Racal in the United Kingdom in 2000, Thales Group has continuously expanded its business, and its civilian business has continued to grow. Now it has developed into a professional electronic high-tech company known for its design, development, and production of aviation, defense, and information technology service products.The Norwegian Institute of Science, Technology and Industry (SINTEF) is headquartered in Trondheim, Norway. It is an independent research organization established in 1950, engaged in contract R&D projects. Based on its R&D in technology, natural sciences, pharmacy, and social sciences, SINTEF provides paid research-based knowledge and related technical services. SINTEF also actively transforms its scientific research results to establish new companies and helps these companies develop. After success, they will sell the shares they own, and the liquidity obtained will then be reinvested in the creation of new knowledge. To ensure a high level of research, SINTEF works closely with the Norwegian University of Science and Technology and the University of Oslo. Many researchers are engaged in formal work in both institutions.The Polish Institute of Electronic Materials Technology (ITME) is a leading multidisciplinary research institution in Poland dedicated to the development of new materials and materials based on innovative tools and components for applications in electronics, microsystems, optoelectronics, micromechanics, and metrology. The high-tech materials, instruments, and components developed by the research institute have been published in many Polish and international journals, which can promote scientific cooperation between it and universities and research institutions and help prospective clients to implement their projects and implement them in industry or use them in the research institutes for short-term continuous production.Leidos, formerly known as Science Applications International Corporation, is a US defense, aerospace, information technology, and biomedical research company headquartered in Reston, Virginia, providing science, engineering, system integration, and technical services. STMicroelectronics is an international semiconductor manufacturer headquartered in Geneva, Switzerland.

impact score than those from the United States and European comparator countries.

Based on the number of academic–corporate nano-publications, the top three academic institutions were all research centers, including the French CNRS, Chinese Academy of Sciences, and the French Alternative Energies and Atomic Energy Commission (CEA) (Fig. 3.7).

Fig. 3.8 illustrates the major worldwide network of academic–corporate collaboration in nanoscience. As a first observation, several large institutions from respective regions in the countries led most of the collaboration efforts. These leading institutions with high academic impact boosted the academic influence of academic–corporate collaborations in these countries. For example, Samsung had the most academic–corporate collaborative nano-publications globally and was also the key entity that led South Korea's cross-sector cooperation

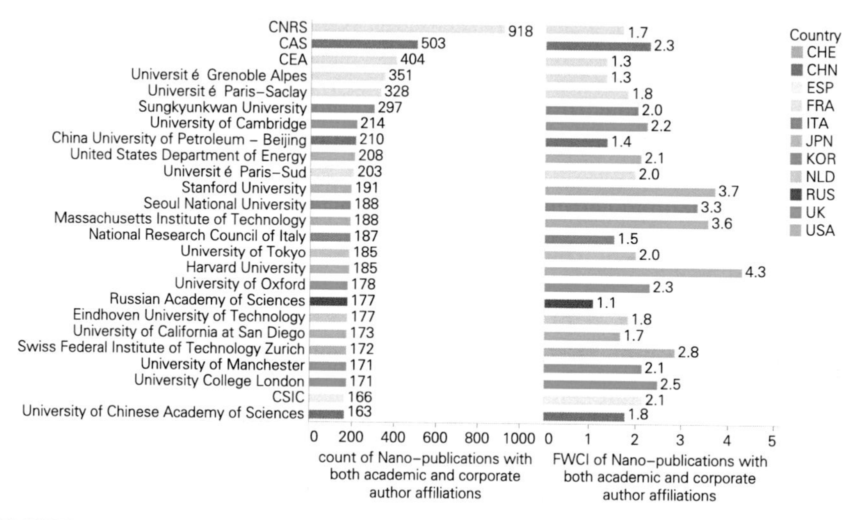

FIGURE 3.7

Top 25 academic affiliations with the most academic–corporate collaborated nano-publications (2015–19). *CHE*, Switzerland; *CHN*, China; *ESP*, Spain; *FRA*, France; *ITA*, Italy; *FWCI*, field-weighted citation impact; *JPN*, Japan; *KOR*, Korea; *NLD*, The Netherlands; *RUS*, Russia; *UK*, United Kingdom; *USA*, United States.

Source: Scopus.

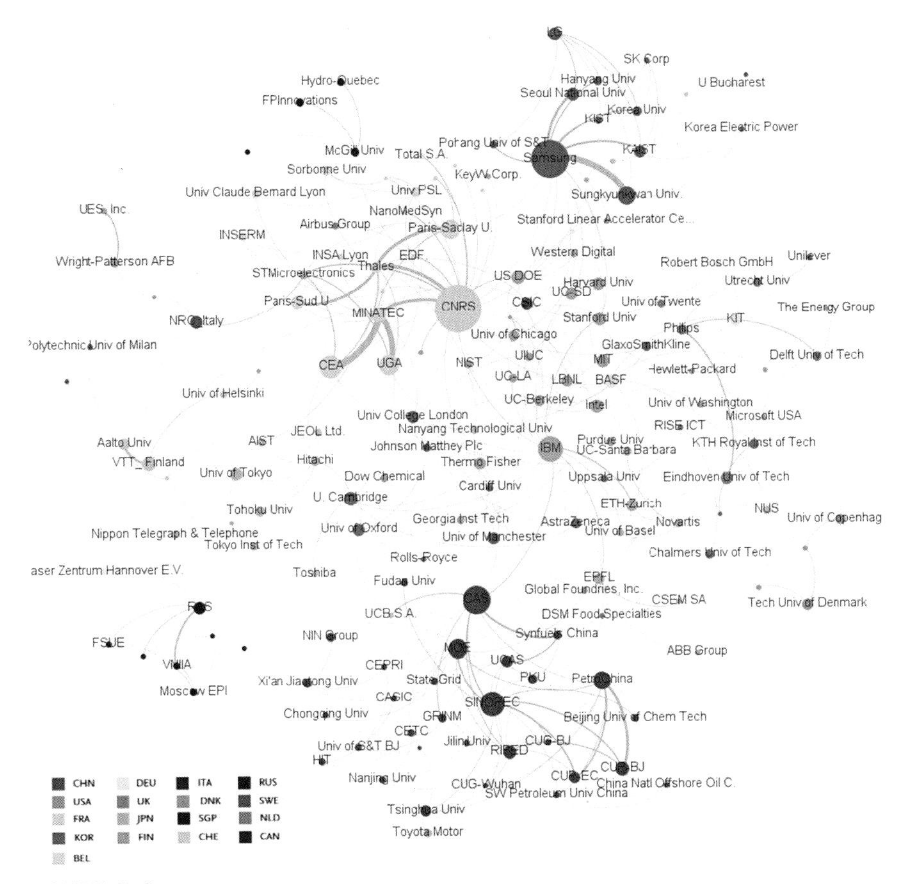

FIGURE 3.8

Collaboration network of the most active institutions with academic–corporate collaborations in the world (2015–19). The bubble size represents the number of publications from academic–corporate collaborations of the given institution. The larger the bubble, the larger the number of publications it represents. Countries are distinguished by bubble colors. The size of the edges represents the number of nano-publications collaborated between the two institutions. The thicker an edge is, the more collaborated nano-publications there are. *BEL*, Belgium; *CAN*, Canada; *CHN*, China; *DEU*, Germany; *DNK*, Denmark; *FIN*, Finland; *FRA*, France; *ITA*, Italy; *JPN*, Japan; *KOR*, Korea; *NLD*, the Netherlands; *RUS*, Russia; *SGP*, Singapore; *SWE*, Sweden; *UK*, United Kingdom; *USA*, United States.

Source: Scopus.

efforts in nanoscience. The United States had many corporate entities participating in academic–corporate collaborations; these cooperation efforts were relatively decentralized and frequent compared to other countries. Besides Samsung, other institutions that had an active role in global nanoscience academic–corporate collaborations including IBM (United States), CNRS (France), MINATEC (France), Sinopec (China), and Chinese Academy of Sciences (CAS) (China).

Second, most collaborating institutions were from the same country as the collaborating corporate entity, indicating a geographic relationship in academic–corporate collaboration. The largest group of academic–corporate collaborators was distributed across the United States, China, France, and South Korea; smaller groups were located in the United Kingdom, Japan, Russia, the Netherlands, Finland, Germany, and Switzerland. A possible explanation for this kind of regional collaboration is that corporate entities might have placed their R&D departments around universities and academic institutes to benefit from academic research, strengthening partnerships with local research centers through geographical proximity. For example, many corporate entities with active academic–corporate collaborations in the United States were located in the northeastern coastal areas, where universities are densely located. Most of China's corporate institutions actively collaborating with academia were in Beijing, where there are substantial academic resources (Fig. 3.5).

In China's nanoscience academic–corporate collaboration network, the CAS collaboration network has produced publications with a relatively high academic impact and extended its partnership to Europe and the United States: for example, with IBM in the United States and UCB Pharma S.A. in Belgium. The collaboration between CAS and IBM was primarily carried out via IBM's Thomas J. Watson Research Center, Albany NanoTech, and other R&D departments at IBM. The CAS Key Laboratory for Biomedical Effects of Nanomaterials and Nanosafety of NCNST and the Institute of High Energy Physics had the most partnerships with IBM's Thomas J. Watson Research Center. Soochow University also participated in many of those studies. In addition, the Institute of Process Engineering of the State Key Lab of Biochemical Engineering, the Beijing Institute of Nanoenergy and Nanosystems, the Wuhan Institute of Physics and Mathematics, the

Institute of Metal Research, the Institute of Applied Chemistry, the Dalian Institute of Chemical Physics, and the Institute of Microelectronics also published nano-publications in collaboration with IBM (Fig. 3.9).

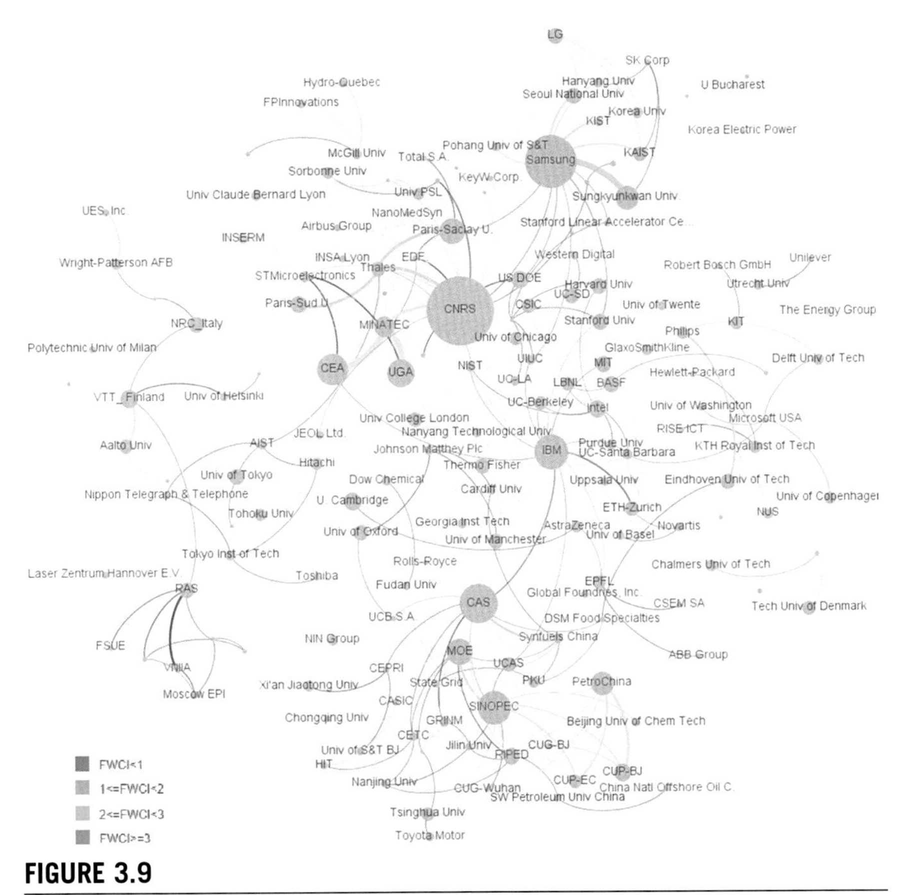

FIGURE 3.9

Network diagram of academic impact of collaborations between the most active institutions in nanoscience in the world (2015–19). The bubble size represents the number of publications from academic–corporate collaborations of this institution. The larger the bubble, the more publications it represents. The tag color represents the sector of the institution, with blue being corporate. The size of edges represents the number of nano-publications collaborated between the two institutions. The thicker an edge is, the more collaborated nano-publications there are. The colors of edges represent the field-weighted citation impact (FWCI) of the collaborated publications; red, yellow, light green, and dark green represent the FWCI from low to high.

Source: Scopus.

3.2 Knowledge transfer: patent citations to academic publications

Basic science is the foundation of research applications, scientific breakthroughs, and technology innovations. Shifting basic research to industrial applications enables scientific developments to serve a wider community, improving the quality of people's daily life. Academic publications that are cited by patents reflect the potential applications for basic research in industry. These patent-cited publications can serve as indicators to help evaluate the transfer of knowledge from basic research to industry, a crucial step for nanoscience.

This section analyzes 66,300 nano-publications cited by patents filed under the five largest intellectual property offices[4], based on Scopus data between 2015 and 2019. The results were used to track the impact of basic nanoscience research on industry.

Share of scholarly output cited by patents

The patent citation rate is the proportion of publications cited by patents in a collection of academic outputs. The rate serves as an indicator to assess an academic research output's possibility of being cited by patents. From 2015 to 2019, 1.04% of nano-publications in the world were cited by at least one patent filed under the five international intellectual property offices (Fig. 3.10). The number was 89% higher than the average global patent citation rate, which was 0.55%. Similar results were found in all key countries (China, Germany, Japan, the United Kingdom, and the United States), indicating that a higher proportion of knowledge from nano-publications was taken up by industry than the global average.

Of all key countries, the United States had the highest nanoscience patent citation rate at 2.4%, which was 2.4 times the rate for all publications combined (1%). The United States, the United Kingdom, Germany, and Japan all had higher patent citation rates for

[4] Patent article citation data is from the EPO, Intellectual Property Office (UK), Japan Patent Office, USPTO, and WIPO.

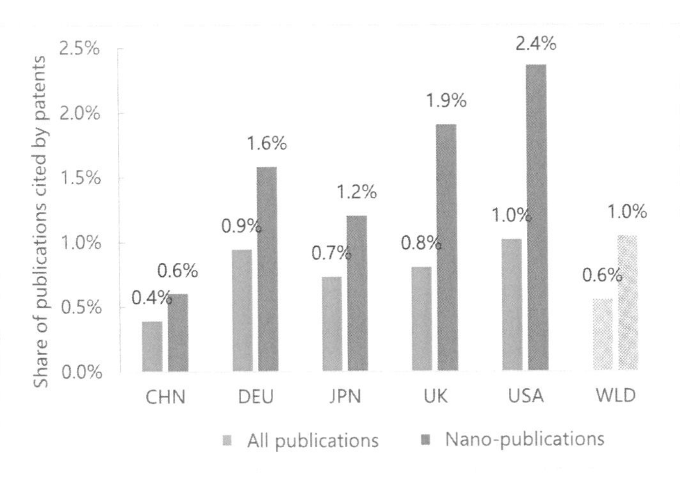

FIGURE 3.10

Share of all publications cited by patents versus share of nano-publications cited by patents (2015–19). *CHN*, China; *DEU*, Germany; *JPN*, Japan; *UK*, United Kingdom; *USA*, United States; *WLD*, world.

Source: Scopus.

nano-publications than the global average. In contrast, China scored lower than the worldwide average for this indicator and was relatively behind the other countries.

Patent citations per 1000 scholarly outputs

The number of patent citations per scholarly output[5] from the five major global international intellectual property offices can serve as an indicator to measure the magnitude of publications cited by patents. The patent citation counts per 1000 nano-publications in the United States and the United Kingdom reached 23.7 and 19.1 between 2015 and 2019 (Fig. 3.11), both higher than their national averages of research in all fields. The results indicated that nano-publications in these two countries had a higher magnitude of patent citations. Meanwhile, the average citation count by patents per 1000 nano-publications in China was 6.0.

[5] Patent citations per scholarly output = Patent Citations Count/Scholarly Output × 1000.

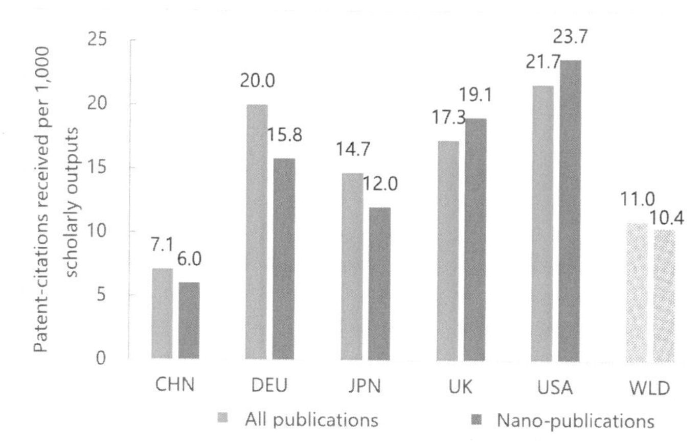

FIGURE 3.11

Patent citations per 1000 scholarly outputs for nano-publications (2015–19). *CHN*, China; *DEU*, Germany; *JPN*, Japan; *UK*, United Kingdom; *USA*, United States; *WLD*, world.

Source: Scopus.

Possible reasons for the relatively low patent citations of nano-publications in China include:

- China's academic–corporate collaborations in nanoscience were far less common than those of other developed countries, negatively influencing the chance of its basic nanoscience research being cited by industry. By analyzing the publications cited by patents, we found that most institutions with high citation counts were those with strong academic–corporate collaborations in nanoscience.
- The citing-patents data came from the five major intellectual property offices: European Patent Office (EPO), the UK's Intellectual Property Office, Japan Patent Office, US Patent and Trademark Office (USPTO), and World Intellectual Property Organization (WIPO). Citations from the Chinese Patent Office were not included, thus reducing the number of patent citations from Chinese literature in the evaluation.

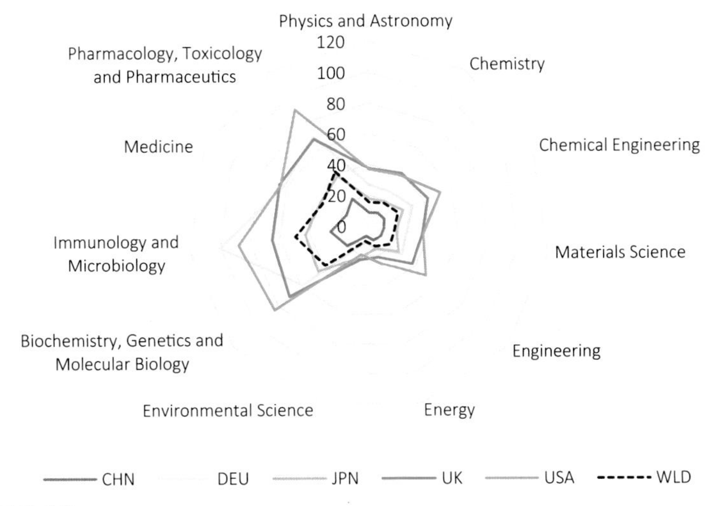

FIGURE 3.12

Patent citations per 1000 scholarly outputs for nano-publications (2015–19). *CHN*, China; *DEU*, Germany; *JPN*, Japan; *UK*, United Kingdom; *USA*, = United States; *WLD*, world.

Source: Scopus.

- Patent citation counts differ across fields. The analysis showed that patent citation counts of nano-publications were higher in the fields of pharmacology, immunology and microbiology, biochemistry, genetics, and molecular biology and the like (Fig. 3.12). However, China's research impact in these fields still lagged behind the United States and the United Kingdom.

Fig. 3.13 lists the top 20 patent owners citing the most nano-publications. More than half of institutions were from Europe or North America. Fig. 3.14 presents the top 20 institutions with the most nano-publications cited by patents.

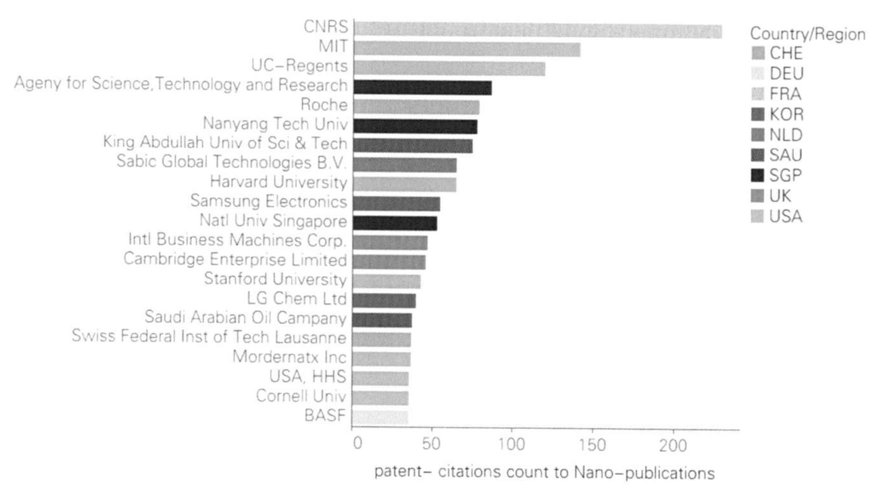

FIGURE 3.13

Top 20 patent owners by patent citation count to nano-publications (2015–19). *CHE*, Switzerland; *DEU*, Germany; *FRA*, France; *KOR*, Korea; *NLD*, the Netherlands; *SAU*, Saudi Arabia; *SGP*, Singapore; *UK*, United Kingdom; *USA*, United States.

Source: Scopus.

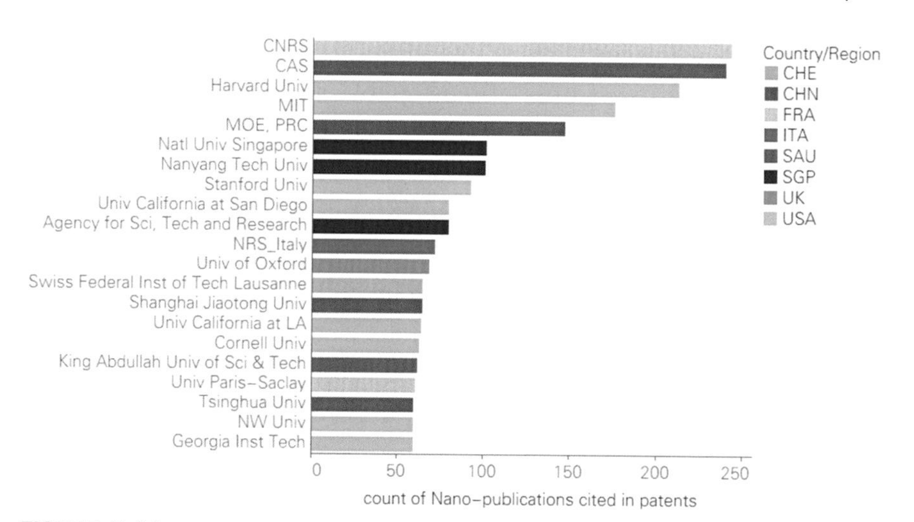

FIGURE 3.14

Top 20 institutions by number of nano-publications cited by patents (2015–19). *CHE*, Switzerland; *CHN*, China; *FRA*, France; *ITA*, Italy; *SAU*, Saudi Arabia; *SGP*, Singapore; *UK*, United Kingdom; *USA*, United States.

Source: Scopus.

3.3 Overview of nano-related patents

Patents are another form of output, reflecting the application prospects of technology in the industry. In this section, nano-related patents are analyzed via patent documents issued by patent licensing agencies in about 115 countries and regions around the world and indexed by PatentSight.[6]

Top inventing countries by nano-related patents

Between 2000 and 2019, there were 693,789 nano-related patents,[7] accounting for 2% of patents worldwide and 72% of all active patents, based on PatentSight data. Of these nano-related patents, 58% were from China, 12% from the United States, and 10% from South Korea. The top five countries with the most nano-related patents accounted for 91% of global nanoscience patents (Fig. 3.15).

The rapidly growing number of nano-related patents in China drove a worldwide surge of patents related to nanoscience. Between 2000 and 2018, the CAGR[8] of global nano-related patents was 14.4%, close to the CAGR of nano-publications in that period. Because the patent data were delayed in 2019, the CAGR was calculated only for 2000–18 (Fig. 3.16).

The indicator used in this book to measure the quality of patents is the Patent Asset Index (PAT), which is derived from a set of three newly developed patent indicators: technology relevance, market coverage, and competitive impact. Technology relevance mainly measures the citations of patents, and market coverage evaluates the degree of protection for a patent or the patent family in the world. Competitive impact is the product of technology relevance and market coverage. The PAT is the aggregation of all patents' competitive impact in a patent collection.

[6] https://go.patentsight.com/.

[7] Nano-related patents: search "Nano*" in the title/abstract/claim of patents. Search time: Mar. 30, 2020. Filing year: 2000–19. For China's patents, we included the utility model patents and deactivated patents; data from Hong Kong, Macau, and Taiwan were omitted.

[8] Affected by the delays in the publication of patent applications, the number of patents in 2019 was excluded when calculating CAGR, because it does not represent the true patent data for 2019.

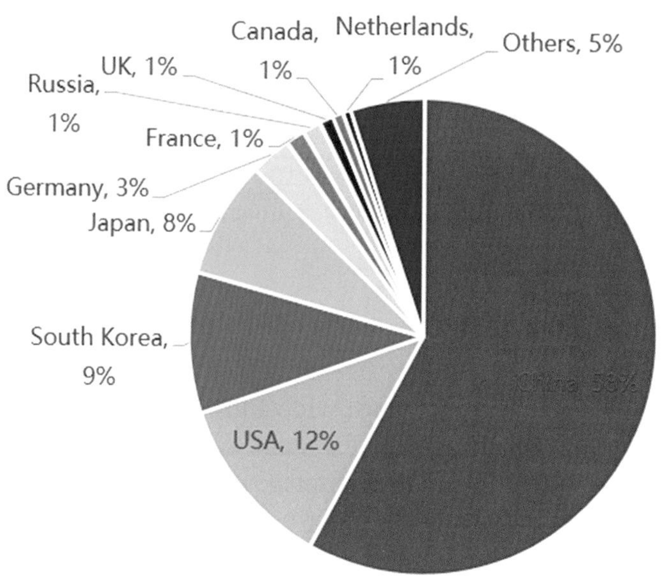

FIGURE 3.15

Share of nano-related patents by top 10 origin countries (2000–19).

Source: PatentSight.

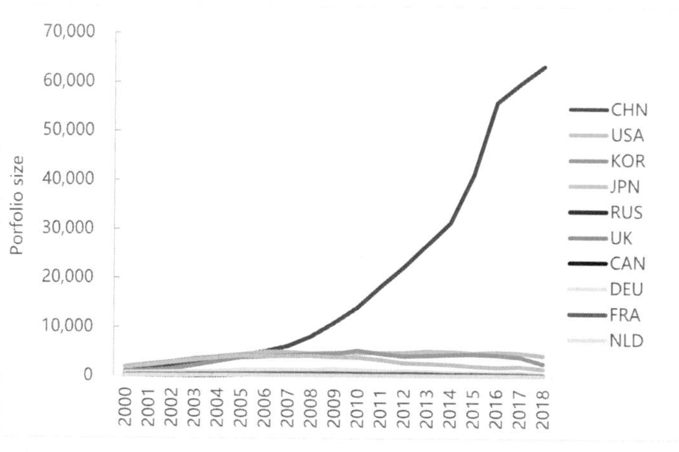

FIGURE 3.16

Trends for nano-related patent counts in top 10 countries of origin (2000–18).[9] *CAN*, Canada; *CHN*, *DEU*, Germany; China; *FRA*, France; *JPN*, Japan; *KOR*, Korea; *NLD*, the Netherlands; *RUS*, Russia; *UK*, United Kingdom; *USA*, United States.

Source: PatentSight.

In general, the more influential a patent is, the higher its quality is considered to be. See Appendix B for detailed descriptions of each indicator.

Although China had a dominant advantage in the total number of nano-related patents, which improved its PAT, its competitive impact ranked low among the top 10 most innovative countries and regions by nanoscience patent count (Fig. 3.17). The lagging gap in its competitive impact comes from its relatively low market coverage and technology relevance. Most of China's nano-related patents were only protected within the nation, resulting in smaller market coverage. Of all China's active nano-related patents between 2000 and 2019, 98% were filed under the National Intellectual Property Administration in China. Compared with China, European countries held a higher

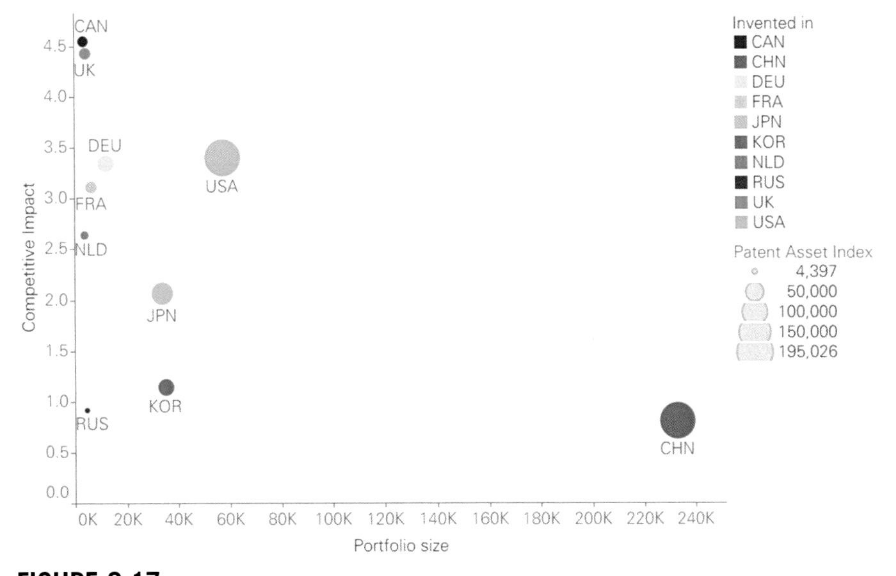

FIGURE 3.17

Top 10 countries of origin by nano-related patents count (2000–19). *CAN*, Canada; *CHN*, China; *DEU*, German; *FRA*, France; *JPN*, Japan; *KOR*, Korea; *NLD*, the Netherlands; *RUS*, Russia; *UK*, United Kingdom; *USA*, United States.

Source: PatentSight.

[9] Because it is affected by the delays in the publication of patent applications, the number of patents in 2019 will be excluded.

Table 3.1 Top 10 countries of origin by nano-related patents count (2000–19).

Invented in	Portfolio size[11]	Patent asset Index	Competitive impact	Technology relevance	Market coverage
China	233,010	188,619	0.8	1.2	0.6
United States	57,364	195,026	3.4	2.1	1.4
Korea	35,079	39,940	1.1	1.2	0.6
Japan	33,912	69,973	2.1	1.6	1.0
Germany	12,314	41,176	3.3	1.9	1.4
France	6,715	20,912	3.1	1.7	1.5
Russia	4,821	4,397	0.9	1.0	0.3
United Kingdom	4,416	19,538	4.4	2.4	1.7
The Netherlands	4,070	10,741	2.6	1.9	1.1
Canada	3,682	16,739	4.5	2.4	1.6

Source: PatentSight.

percentage of patents in their own patent offices, and more patents from these countries were also filed under the USPTO. France, Germany, and the United Kingdom had 61%, 57%, and 70% of nano-related patents filed for protection under the USPTO. The United Kingdom's and Canada's nano-related patents had relatively more protection from the EPO and the WIPO. For example, 11% and 10% of nano-related patents in the United Kingdom and Canada, respectively, were WIPO patents (Table 3.1).[10]

Top 10 patent owners per Patent Asset Index of nano-related patents

The patent owner is the entity that owns the patent rights. Table 3.2 and Fig. 3.17 list the top 10 patent owners in industry and academia by their PAT of nano-related patents.

---

[10] Data source: PatentSight.

[11] To ensure the accuracy of the PAT, this table counts only the number of valid patents and does not include Chinese utility model patents.

Table 3.2 Patent metrics of top 10 patent-owning corporates by Patent Asset Index of nano-related patents (2000–19).

Owner (corporate)	Patent asset Index	Portfolio size	Competitive impact	Technology relevance	Market coverage
Samsung	10,738	4,153	2.6	1.9	1.2
LGChem	4,681	1,744	2.7	1.6	1.2
Foxconn	4,532	3,070	1.5	1.2	1.0
Intel	3,427	1,189	2.9	1.9	1.4
BASF	3,253	708	4.6	2.4	1.7
Dow Inc	2,885	704	4.1	2.2	1.7
Fujifilm	2,808	1,124	2.5	2.0	1.1
Merck KGaA	2,764	526	5.3	3.1	1.7
Johnson & Johnson	2,666	339	7.9	3.5	2.0
BOE	2,645	1,278	2.1	1.7	1.2

From PatentSight.

Many patent-owning companies have departments actively involved in nano-related academic–corporate collaborations. For example, Samsung and LG participated the most in academic–corporate collaborations in South Korea, and the two enterprises produced a higher patent output. Sinopec from China ranked third for patent counts, but its competitive impact was relatively low, and its PAT did not make it into the top 10. The United States had four companies among the top 10 patent owners, the most country. Intel had the highest nano-related PAT, followed by Dow Chemistry, Johnson & Johnson, and Boeing. Despite being actively involved in academic–corporate collaboration in the United States, IBM's PAT for nano-related patents did not make its way to the top 10. This was because 60% of IBM's nano-related patents were filed in the last 5 years, and the patents' technology relevance had yet to reach its full potential. Between 2000 and 2019, 45% of Foxconn's nano-related patents shared co-ownership with Tsinghua University, demonstrating close cooperation between the two entities. BASF and Merck, two German companies, had nano-related patents with a relatively higher competitive impact, ranking in the top three among all 10 companies (Table 3.2, Fig. 3.18).

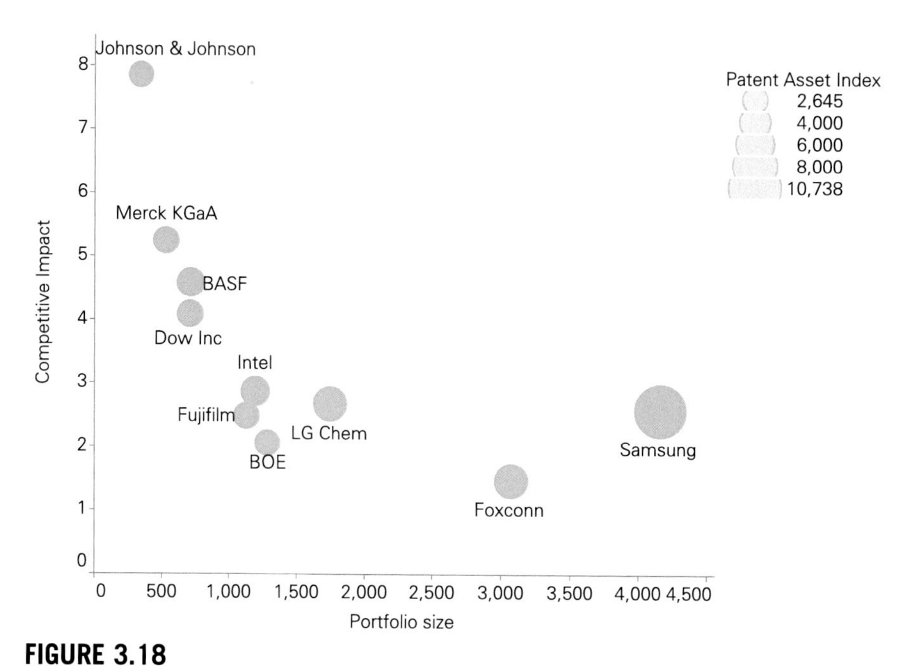

FIGURE 3.18

Top 10 patent-owning corporates by Patent Asset Index of nano-related patents (2000–19). *DEU*, Germany; *JPN*, Japan; *KOR*, Korea; *TWN*, Taiwan; *USA*, United States.

Source: PatentSight.

From 2000–19, four of the top 10 patent-owning academic institutions ranked by the PAT of nano-related patents were from China. Although China had an edge in the number of patents, its competitive impact was relatively low. CAS had the most nano-related patents, but its market coverage was small because most patents were protected only in China. Academic institutions from the United States, such as MIT, Harvard University, the University of California, and Broad Institute,[12] had higher competitive impacts. Among these US institutions, Broad Institute only had 43 nano-related patents between 2000 and 2019, but its competitive impact was immense, especially for technology relevance, indicating that the patents were cited by other patents more often (Table 3.3, Fig. 3.19).

[12] Broad Institute is a research collaboration center owned by both MIT and Harvard University.

Table 3.3 Patent metrics of top 10 academic institutions with patent ownership by Patent Asset Index of nano-related patents (2000–19).

Owner (academic)	Patent asset Index	Portfolio size	Competitive impact	Technology relevance	Market coverage
Chinese Academy of Sciences	10,795	11,747	0.9	1.4	0.6
Tsinghua University	4,847	3,220	1.5	1.5	1.0
MIT	4,717	711	6.6	3.5	1.5
Harvard University	3,333	364	9.2	4.6	1.7
University of California	3,331	1,265	2.6	1.8	1.4
Semiconductor Energy Lab	2,185	449	4.9	3.0	1.4
Broad Institute	1,980	43	46.0	18.7	2.3
Zhejiang University	1,851	2,061	0.9	1.4	0.6
Southern University of Science and Technology	1,796	1,859	1.0	1.6	0.6
CNRS	1,689	902	1.9	1.1	1.6

Source: PatentSight.

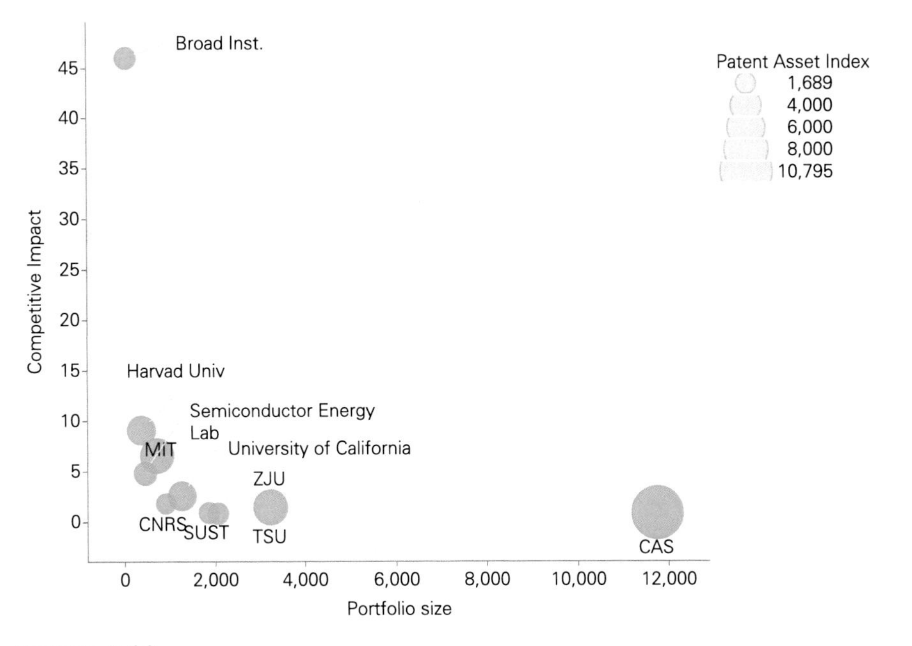

FIGURE 3.19

Top 10 academic institutions with patent ownership by Patent Asset Index of nano-related patents (2000–19). *CHN*, China; *FRA*, France; *JPN*, Japan; *USA*, United States.

Source: PatentSight.

Patent transfer

Patent ownership transfer

In this section, we analyze the nano-related patents filed in China, the United States, Germany, the United Kingdom, and Japan that went through a change in patent ownership. The reassignors and reassignees[13] are analyzed to understand the process of patent assignment from academic institutions to industries.

Between 2015 and 2019, a total of 976 nano-related patents were transferred from research institutes and universities to corporate entities. Of these assigned patents, 415 were from the United States, 306 were from China, and 166 were from Japan, accounting for 43%, 31%, and 17%, respectively, of transfers. In addition, 60 transferred patents were from Germany and 51 were from the United Kingdom.

Tsinghua University was the biggest patent reassignor, with the largest number of patent assignments. Between 2015 and 2019, the university assigned 217 nano-related patents to Hon Hai Technology Group (Foxconn). These patents were mainly related to nanotubes, capacitive touch screens, semiconductors, storage, and graphene sheets. Twelve nano-related patents were assigned from the California Institute of Technology to Samsung (Tables 3.4 and 3.5).

Joint ownership

Besides patent assignment, academia and industry can jointly own patent rights. Table 3.6 lists academic and corporate partners worldwide that co-owned[14] the most nano-related patents between 2015 and 2019. Wuhan University and Wuhan Fanglue Digital Technology jointly owned 433 nano-related patents. Among the patents, those

---

[13] Here, reassignor represents an academic entity and reassignee represents a corporate entity. Data include deactivated patents and exclude utility model patents. Filing Date: Jan. 1, 2015 to Dec. 31, 2019, Title/Abstract/claim=(Nano*), Invented In=(CN, DE, GB, JP, US).

[14] Both entities own the patent. Data include deactivated patents and exclude utility model patents.

Table 3.4 Top 10 patent reassignors with nano-related patents transferring from the academic to the corporate sector (2015–19).

Top 10 academic reassignor	Count of reassigned patents (2015–19)
Tsinghua University	217
University of California	31
Georgia Institute of Technology	24
University of Tokyo	16
Japan Institute of Industrial Technology (AIST)	16
California Institute of Technology	15
Massachusetts Institute of Technology	15
Osaka University	14
Kyoto University	11
Harvard University	10
Tohoku University	10

Source: PatentSight.

Table 3.5 Top 10 patent reassignees with nano-related patents transferring from academic to corporate sector (2015–19).

Top 10 reassignees (company)	Count of reassigned patents (2015–19)
Hon Hai Precision Industry Co., Ltd., Taiwan	215
Samsung Electronics Co.	29
Toyota Jidosha Kabushiki Kaisha, Japan	13
UT-Battelle, LLC, Tennessee	12
Proctor & Gamble Company, Ohio	11
International Business Machines Corporation, New York	10
Zeon Corporation, Japan	9
Alliance for Sustainable Energy, LLC, Colorado	8
BASF	7
Rohm & Haas Elect Materials LLC	7

Source: PatentSight.

Table 3.6 Institution pairs with the most nano-related patents jointly owned by academic and corporate sectors (2015–19).

Co-owner of academic entity	Co-owner of corporate entity	Count of co-owned patents (2015–19)
Wuhan University	Wuhan Wuda Fang Lue Digital Technology Co. Ltd.	433
Tsinghua University	Foxconn	254
Hebrew University	Yissum R&D (in: Hebrew University)	67
University of Oxford	Oxford Innovation (in University of Oxford)	57
Soochow University	Jiangsu Industrial Technology Research Institute	49
Soochow University	SVG group	47
CNRS	INSERM	43
University of Tennessee	Battelle	39
Wuyi University	Jinjiang Ruibi Tech	34
South China Normal University, Shenzhen Guohua Photoelectric Tech Co. Ltd.	Shenzhen Guohua Photoelectric Research Institute	30

Source: PatentSight.

with the highest PATs were in the technology fields of physics > optics > resonator; physics > chemical treatment > photocatalyst; physics > optics > antireflection film; electronics > electricity > electrode materials; chemistry > nanotechnology > semiconductor; and information > analytical materials > electrochemical sensor.

CHAPTER 4

Factors that promote the development of nanoscience

Key findings

132,220 awards were related to nanoscience, accounting for approximately 3.6% of all awards, with a compound annual growth rate (CAGR) of 3% (2009–18).

Material science, chemistry, and physics and astronomy had the highest ratio of awards related to nanoscience (29%, 17.9%, and 14.8% respectively). The number of nano-related awards also ranked first in materials science (2009–18).

A higher CAGR for nano-related awards in eight subjects: The CAGR for nano-related awards in eight analyzed subjects[1] in

Pharmacology, Toxicology and Pharmaceutics, and Energy
Nano-related awards in pharmacology, toxicology, and

[1] Only pharmacology, toxicology, and pharmaceutics is an exception. That is because the CAGR of all awards in the subject pharmacology, toxicology, and pharmaceutics is high (14% in the book). However, the CAGR of nano-related awards in pharmacology, toxicology, and pharmaceutics is the highest among all subjects (4.9%).

Big Data Analysis of Nanoscience Bibliometrics, Patent, and Funding Data (2000–2019)
https://doi.org/10.1016/B978-0-323-91311-9.00004-9

this chapter is higher than the respective subject's average CAGR, including biochemistry, genetics, and molecular biology; chemical engineering; chemistry; energy; engineering; environmental science; materials science; medicine; and physics and astronomy.

pharmaceutics and in energy had the fastest growth (the highest CAGR) among all the subjects; their CAGR was 4.9% (2009–18).

The National Natural science Foundation of China (NSFC) funded the largest number of nano-related awards. Between 2009 and 18, the NSFC funded 27,387 nano-related awards, accounting for 8.7% of all of their awards.

The National Science Foundation (NSF) funded the highest amount of nano-related awards. Between 2009 and 18, the NSF funded around US $6.9 billion for 12,942 nano-related awards, ranking number one among all funders.

28%

In 2019%, 28% of global nano-publications were international collaboration efforts. The percentage had increased 8 points since 2000 and was currently higher than the global average.

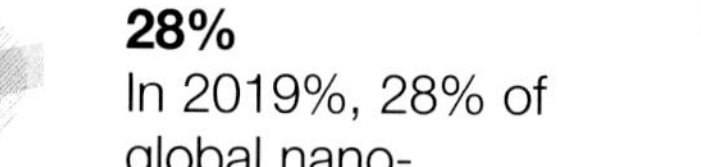

2.5

China's international collaboration effort in nanoscience continued to grow. Its internationally collaborated nano-publications had the highest academic influence of all comparator countries, with a field-weighted citation impact of 2.5 (2010–19).

4.1 Funding analysis for nanoscience

Research funding has become a crucial means and channel for society to invest in science and technology, and it is also a crucial step for scientific advancement. This section focuses on the funding for nanoscience-related projects in all granted awards, to explore the driving forces behind scientific research development.

Elsevier's funding data platform, Funding Institutional, provides funding information about more than three million projects, covering 3500 government and private fundraising organizations worldwide. See Appendix C for further details about the platform. This book provides an assessment of nano-related projects worldwide based on information from Funding Institutional.

Funding information collected by the platform covered multiple global funding agencies, including information about more than 2000 funding agencies in the United States, such as the National Science Foundation (NSF), National Institutes of Health (NIH), NASA, Department of Energy, and Department of Defense; projects from the National Natural Science Foundation of China (NSFC); projects funded by the German Science Foundation (Deutsche Forschungsgemeinschaft [DFG]), the Federal Ministry for Food and Agriculture (Bundesministerium für Ernährung und Landwirtschaft), Volkswagen Foundation, and the German Environment Agency; projects funded by 729 foundations in the UK (including UK Research and Innovation, Wellcome Trust, etc.); and projects from the Japan Society for the Promotion of Science (JSPS). Funding agencies from South Korea have not yet been included in this platform.

Number and value of nano-related awards

Between 2009 and 2018, a total of 132,220 awards[2] were related to nanoscience, accounting for approximately 3.6% of global awards, according to Funding Institutional. The total rose from 3% in 2009 to 4% in 2018, resulting in a compound annual growth rate (CAGR)

[2] Nano-related awards: With "Nano" in the title or abstract of the award. Search date: Mar. 2020. The 2019 data were not fully updated. Thus, 2009–18 was selected for funding analysis.

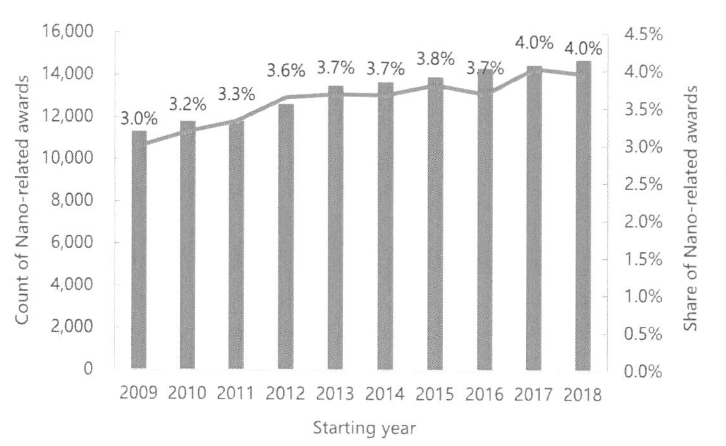

FIGURE 4.1

Trends for the share and number of nano-related awards in the world (2009–18).

Source: Funding Institutional.

of 3% (Fig. 4.1). In that period, the number of awards worldwide remained mostly unchanged. Based on available data, the total amount of funding for nano-related projects reached US $42.3 billion[3] between 2009 and 2018.

Nano-related awards in each subject

Fig. 4.2 illustrates the number and growth of nano-related awards in key subjects with the highest number of nano-related awards, including biochemistry, genetics, and molecular biology; chemical engineering; chemistry; energy; engineering; environmental science; materials science; medicine; pharmacology, toxicology, and pharmaceutics; and physics and astronomy. Among them, nano-related awards in materials science, physics and astronomy, and chemistry had the highest percentages. The results aligned with subjects' ranking by share of nano-publications, indicating a positive correlation between funding and academic output (Fig. 4.3).

[3] A small amount of budget information was unavailable in Funding Institutional.

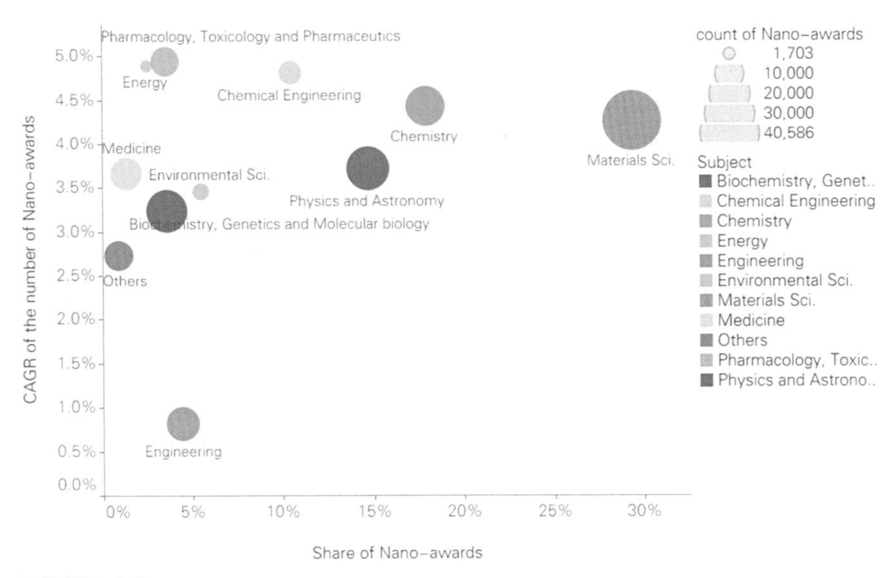

FIGURE 4.2

Number, share, and compound annual growth rate (CAGR) of nano-related awards in each discipline in the world (2009–18).

Source: Funding Institutional.

Materials science has the largest number and share of nano-related awards. Between 2009 and 18, a total of 40,586 funded projects in materials science were nanoscience related, accounting for 29% of awards in the field with a CAGR of 4.3%. The average CAGR for all awards in materials science was 2.5%. Most nano-related funded projects were in the subject of chemistry (17,903 projects, at 17.9% and a CAGR of 4.4%), followed by physics and astronomy (21,543 projects with a share of 14.8% and a CAGR of 3.7%).

Nano-related awards in energy and in pharmacology, toxicology, and pharmaceutics had the fastest growth with the highest CAGR. The exponential growth of awards in pharmacology, toxicology, and pharmaceutics as a whole drove the surge in numbers of nano-related awards. The CAGR of nano-related awards in pharmacology, toxicology, and pharmaceutics was 4.9%, whereas this subject's overall CAGR in awards was 14%. Except for those in pharmacology, toxicology, and pharmaceutics, the growth rates of nano-related awards in each subject, as measured by CAGR, were higher than the subjects' overall averages.

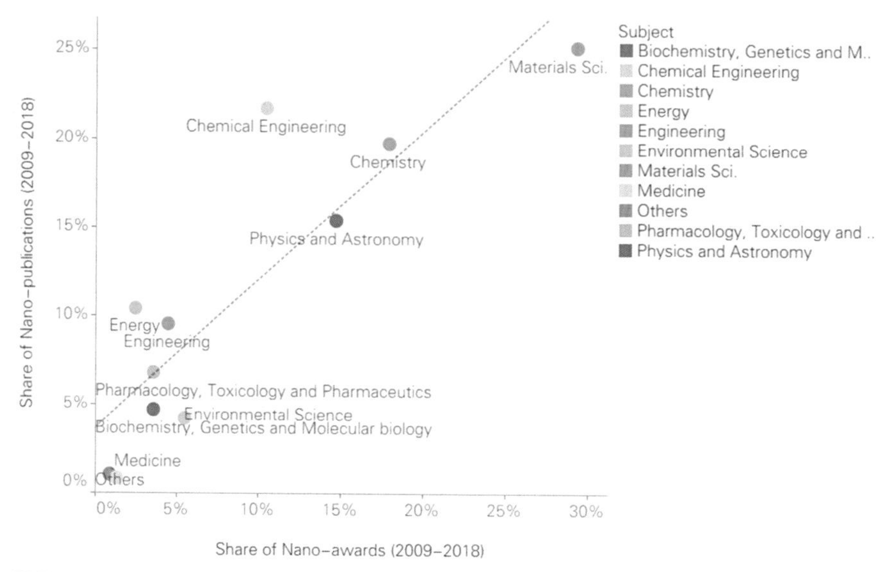

FIGURE 4.3

The share of nano-related awards (x-axis) versus the share of nano-publications (y-axis) in each subject (2009–18).

Source: Funding Institutional, Scopus.

Energy placed second among the key subjects discussed subsequently, which are ranked by the CAGR of their nano-related awards. The growth of nano-related awards in energy was faster than its average of all awards. The CAGR of nano-related awards in energy was 4.9%, whereas the CAGR of all awards in the subject was −3.3%.

Top 10 funders by count of nano-related awards

Fig. 4.4 lists the top 10 institutions that funded the most nano-awards from 2009 to 2018. These institutions funded 64% of global nano-related awards in that period. Among them, the NSFC funded the most nano-related awards, with a total of 27,387, accounting for 8.7% of all NSFC-funded awards. The number of nano-related awards funded by the NSFC also continued to grow, and its growth rate (a CAGR of 17.8%) was the fastest among the top 10 institutions.

The funding trends in nano-related awards differed across major funders in the United States. With 10.1% of its awards relevant to nanoscience, the NSF had the highest share of nano-related awards.

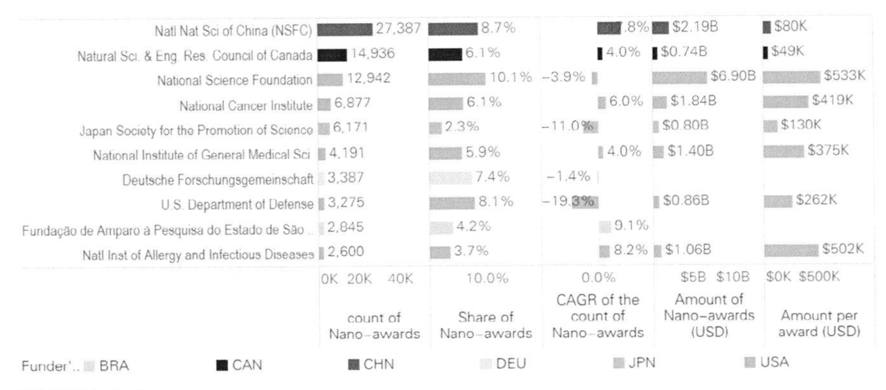

FIGURE 4.4

Top 10 funders[4] that funded the most nano-related awards in the world (2009–18). *BRA*, Brazil; *CAN*, Canada; *CHN*, China; *DEU*, Germany; *JPN*, Japan; *USA*, United States.

Source: Funding Institutional.

The number of nano-related awards from the US Department of Defense declined the fastest, with a CAGR of −19.3%. However, funding continued to grow for nano-related awards sponsored by the three funders affiliated with the NIH: the National Cancer Institute, the National Institute of General Medical Science, and the National Institute of Allergy and Infectious Diseases.

A comparison of the number of nano-related awards (Fig. 4.4) and academic output (Fig. 4.5) per leading funders showed that many funding agencies overlapped in both lists, including the NSFC, NSF, Foundation for the NIH, German Science Foundation, and JSPS. This indicated a positive correlation between the investment of funding agencies and academic outputs. The comparison excluded funders whose data were unavailable in Funding Institutional, which did not cover funding information other than the NSFC for Chinese funders and does not yet include funding data from South Korea.

[4] The project amount information for DFG and the Brazil Foundation is unavailable, and the funding agency data from South Korea are not yet included on this platform.

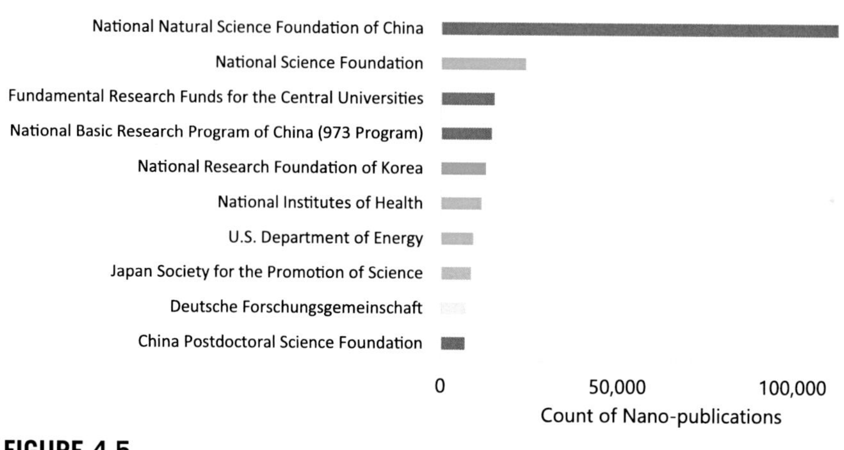

FIGURE 4.5

Top 10 funders by nano-publication count (2009–18).

Source: Scopus.

4.2 International collaboration in nanoscience

As globalization increases, international collaborations in the scientific research community have become more frequent. Cooperation across institutions, regions, continents, and time zones has brought about the exchange and sharing of knowledge, which has increased academic impact. For example, internationally collaborated publications from 2010 to 2019 that are indexed in the Scopus database have an average count of citations per publication 1.68 times the overall average.

International collaboration rate in nanoscience compared with all research fields combined

As one of the key areas in modern science, nano-related research showed a higher degree of global cooperation. Between 2010 and 2019, 21% of the world's publications across all research fields were published under international collaboration efforts, but this number was 25% in nano-related research (Fig. 4.6). In all comparator countries, the international collaboration rate of nano-publications was higher than the nation's average and the world, indicating that global cooperation became more frequent in nano-related research activities.

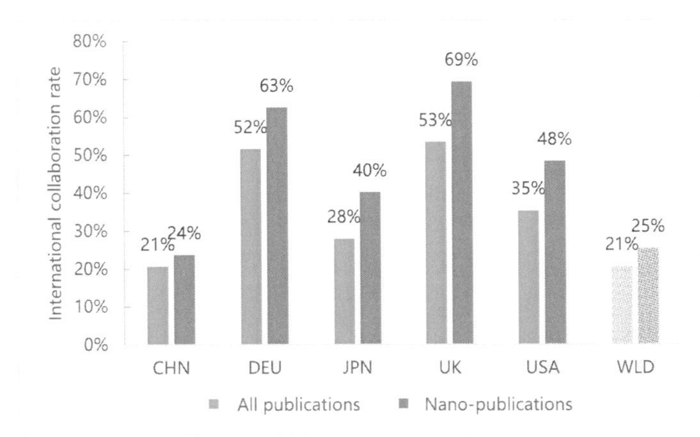

FIGURE 4.6

International collaboration rates in nanoscience and all fields for comparator countries and the world (2010–19). *CHN*, China; *DEU*, Germany; *JPN*, Japan; *UK*, United Kingdom; *USA*, United States; *WLD*, world.

Source: Scopus.

China's international collaboration effort still lags behind that of other developed countries. Between 2010 and 2019, 24% of nano-publications were produced with global cooperation, slightly lower than the global average of 25% and lower than the comparator countries. There are two possible reasons for this. First, China contributed to 39% of worldwide nano-publications in 2019, becoming the country with the highest nano-related academic output. Having many Chinese academic institutions that led the world in nano-related research created a path for a more robust domestic collaboration than global cooperation. As shown in Fig. 4.8, China's domestic collaboration rate continued to outperform its international collaboration rate. Second, compared with developed countries in Europe and the United States, China had a relatively low global collaboration rate in overall scientific research, affecting nanoscience's international collaboration.

Academic impact of China's internationally collaborated nano-publications

International collaboration has improved the overall academic impact of nano-publications. Between 2010 and 2019, the field-weighted citation impact (FWCI) of internationally collaborated nano-publications in the world was 1.9, which was 26% higher than the FWCI for all nano-publications (Fig. 4.7). The same observation was found in all key countries, indicating that international collaboration is beneficial for improving academic impact.

In addition, internationally collaborated nano-publications with China as a partner had a higher academic impact compared with other countries. Between 2010 and 2019, the FWCI of internationally collaborated nano-publications in which China participated was the highest among all comparators (FWCI = 2.5), which was higher than publications in which the United States participated (FWCI = 2.3). After China, the FWCI of internationally collaborated nano-publications

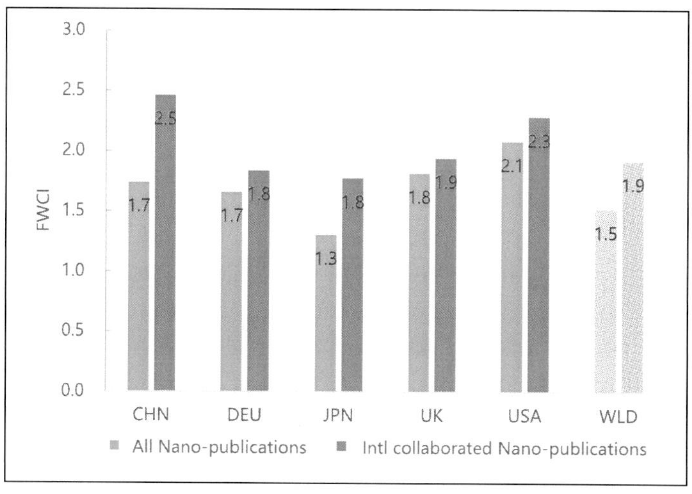

FIGURE 4.7

Comparison of the field-weighted citation impact (FWCI) of internationally collaborated nano-publications and of nano-publications in each comparator country and the world (2010–19). *CHN*, China; *DEU*, Germany; *JPN*, Japan; *UK*, United Kingdom; *USA*, United States; *WLD*, world.

Source: Scopus.

with Japan as a partner was 1.8, which was 36% higher than the overall average of the nation's nano-publications (FWCI = 1.3).

Upward trend in international collaborations in nanoscience

The international collaboration rate in nanoscience increased significantly in key countries and the world in the past few decades (Fig. 4.8). Globally, the international collaboration rate of nano-publications in the world increased from 18% in 2000 to 26% in 2019. Only the institutional collaboration share showed a downward trend, indicating that the geographical footprint of cooperation in nano-related research was expanding.

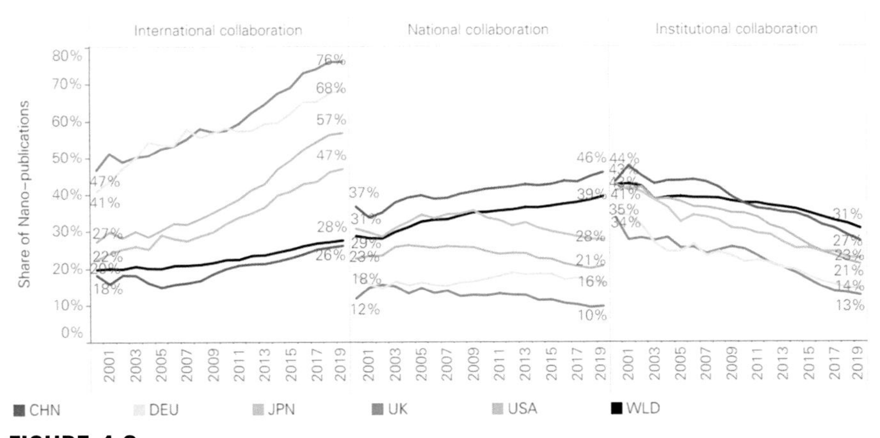

FIGURE 4.8

Trend in nano-publications' share with different types of collaborations in each comparator country and the world (2000–19). *CHN*, China; *DEU*, Germany; *JPN*, Japan; *UK*, United Kingdom; *USA*, United States; *WLD*, world.

Source: Scopus.

Conclusions

Nanoscience and nanotechnology progressed rapidly between 2000 and 2019, with growth rates in nano-related academic output and patent output significantly higher than the overall world average. China was an important driver of this growth of global nanoscience output. In this book, we use the field-weighted citation impact (FWCI), which is the normalized citations, as a proxy for academic impact. China's academic impact in nanoscience was also increasing: its FWCI increased from 1.3 in 2000 to 1.9 in 2019, with the same trend shown for its share of the 1% most highly cited papers in nanoscience. In 2000–19, for every four Chinese 1% most highly cited publications, one was relevant to nanoscience. As shown in Fig. 1, China ranked highly for many key indicators in nanoscience.

As a universal science, nanoscience and nanotechnology stood at the intersection between various basic sciences and were integrated with multiple research areas, enhancing academic output and the impact of scientific research. Nanoscience had the closest relationship with physical sciences, which includes subjects such as materials, chemistry, chemical engineering, physics and astronomy, and engineering. However, emerging trends also showed a strengthened integration of nanoscience with life sciences and health sciences. In recent years, the percentage of research outputs involving nanoscience has increased in biochemistry, genetics and molecular biology, and pharmacology, toxicology, and pharmaceutics. Furthermore, nanotechnology has been significant to each fundamental science for its contribution to overall academic output and also indispensable to some of the most prominent research topics.

Besides being imperative to basic science, nanoscience and nanotechnology have contributed immensely to industry via advanced technology application. In recent years, many countries have introduced policies to promote the transfer of scientific research results to

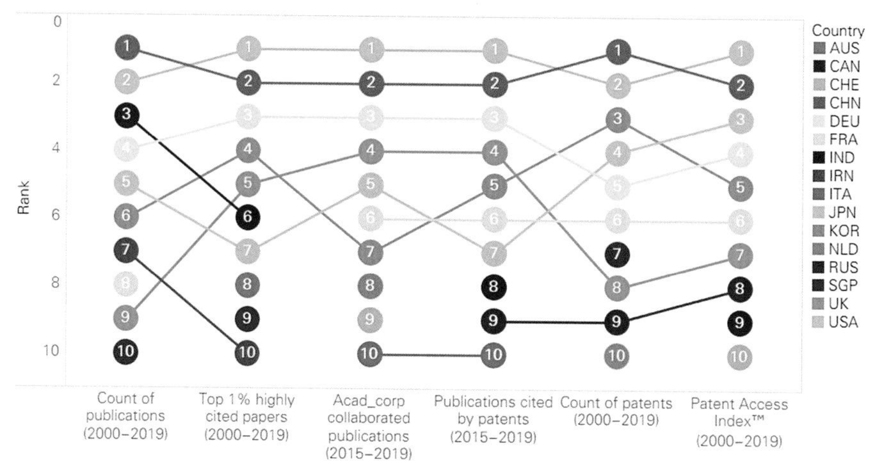

FIGURE 1

Top 10 countries by main evaluation indicators in nanoscience. *AUS*, Austria; *CAN*, Canada; *CHE*, Switzerland; *CHN*, China; *DEU*, Germany; *FRA*, France; *IND*, India; *IRN*, Ireland; *ITA*, Italy; *JPN*, Japan; *KOR*, Korea; *NLD*, the Netherlands; *RUS*, Russia; *SGP*, Singapore; *UK*, United Kingdom; *USA*, United States.

From Scopus, PatentSight.

industrial applications, and the collaboration between industry and academia has opened doors to this opportunity. Although China's academic–corporate collaboration in nanoscience still lagged behind other developed countries, its academic–corporate collaborated nano-publications have proliferated, showing the highest growth rate among comparator countries during the study period. The United States, Germany, the United Kingdom, and Japan had relatively higher academic–corporate collaboration rates in nanoscience, all of which were higher than their national averages. Lessons can be learned from accumulated experience in academic–corporate collaborations in these countries. The United States, France, South Korea, and China had the best global academic–corporate collaboration network in nanoscience. The collaborating corporate entities in China were mainly from the petrochemical industry, and the academic impact of these partnerships still has room for improvement. Corporate entities involved in academic–corporate collaboration from the United States were mostly high-tech and research and development–intensive companies such as IBM, whose academic output had a more noteworthy impact.

The citation of academic publications by patents reflects the uptake of basic research results by industrial applications and indicates the economic benefits of academic research. Compared with other key countries, China was behind in patent citations received per 1000 publications in nanoscience for three main reasons. First, part of the lag resulted from varying disciplinary focus from different countries, in which those with publications in life sciences and medicine can have higher patent citations. Second, the study included only patent citation data from the World Intellectual Property Organization, US Patent and Trademark Office, European Patent Office UK Patent Office, and Japan Patent Office, resulting in a limited data set that can contribute to lower patent citations. Third, the analysis showed that corporate entities active in nano-related fields actively cited nano-publications, indicating academic–corporate collaboration closely related to the number of patent-cited nano-publications. From this perspective, China also needs to improve its academic–corporate collaboration in nanoscience.

Over the past 2 decades, the application of nanotechnology has boomed in the industrial sector, with a total of more than 690,000 nano-related patents worldwide, 58% of which came from China. China's nano-related patents had an absolute advantage in number but had a relatively low impact owing to the patents' low market coverage and technology relevance. Historically, most Chinese patents applied for protection only in China, whereas those in Europe and America applied for protection across different countries and regions.

More funding was available for nano-related research projects between 2009 and 2018, and the total of nano-related awards was increasing in both number and percentage, pushing development in nanoscience. Comparison between top funders by the amount of nano-awards and top funders by the scholarly output of nano-research showed a positive correlation between funding strength and academic output: a good return on investment.

Financial support and the development of technology have promoted frequent academic exchange. Cross-regional collaboration can also inspire ideas and facilitate the communication and spread of knowledge, which is conducive to advancing science. As found by this book, nano-research is undergoing globalization. The international collaboration rate of nano-publications was higher than the global average, and global cooperation has positively affected academic impact in many countries. China has had a remarkable academic impact resulting from its international collaboration in nanoscience and has become a competitive candidate as a global partner.

Postscript

This book was written by the National Center for Nanoscience and Technology, China and Elsevier's Analytical Services team. This book used quantitative research methods to analyze, evaluate, and compare the impact of output from nanoscience and nanotechnology in key countries and the world to understand current development trends in the field.

This book used bibliometric methods, such as the number of publications, the field-weighted citation impact, the academic–corporate collaboration rate, and the patent citation rate. These indicators are limited by the performance of scientific research and the metric-determining methodologies. The impact of external factors can also contribute to systemic biases. Over the past decade, the best practices in the field of bibliometrics indicators have guided the interpretation of indicator analysis results and revealed factors that should be used for specific analyses and evaluations. The analysis methods in this book were derived from these best practices, informed by experts' insight, and based on a large quantity of literature and monographs. Although the book may have limitations owing to the selection of indicators and their analysis, it provides a big-picture view of nanoscience and nanotechnology in the global landscape.

The analysis in this book was based on 1.42 million academic publications and over 600,000 patents. However, there are also flaws in the book's data, such as queries that do not cover all relevant works in nano-related research and incomplete international scientific funding data. Although not fully comprehensive, the scale of the data has ensured that the book reflects the real-world situation.

Finally, owing to time constraints in completing this study, errors and omissions are inevitable in the book. We appreciate your understanding and feedback.

Tables

Table A.1

Table A.1 Top 100 corporate entities by academic–corporate collaborative nano-publications (2015–19).

Corporate name	Country code	Academic–corporate collaborative nano-publications	Field-weighted citation impact of academic–corporate collaborative nano-publications
Samsung	KOR	697	1.86
IBM	USA	449	2.17
SINOPEC	CHN	428	1.42
PetroChina	CHN	290	1.21
MINATEC	FRA	252	1.19
VTT Technical Research Centre of Finland Ltd.	FIN	220	1.35
Intel	USA	208	1.58
BASF	DEU	194	2.16
Research Institute of Petroleum Exploration and Development	CHN	194	1.47
LG Corporation	KOR	186	1.37
Thermo Fisher Scientific	USA	174	2.20
Thales	FRA	167	1.95
SINTEF	NOR	158	1.32
Institute of Electronic Materials Technology	POL	141	0.94
Hitachi	JPN	132	1.01

Continued

Table A.1 Top 100 corporate entities by academic–corporate collaborative nano-publications (2015–19).—cont'd

Corporate name	Country code	Academic–corporate collaborative nano-publications	Field-weighted citation impact of academic–corporate collaborative nano-publications
Koninklijke Philips N.V.	NLD	131	1.74
AstraZeneca	UK	127	1.88
GlaxoSmithKline	UK	118	1.77
STMicroelectronics	CHE	109	0.97
Leidos Inc	USA	108	1.98
Northwest Institute for Nonferrous Metal Research	CHN	108	1.29
Synfuels China Co., Ltd.	CHN	101	2.07
General Motors	USA	97	2.64
General Research Institute For Nonferrous Metals	CHN	96	1.13
Dow Chemical	USA	95	1.73
FP Innovations	CAN	92	1.91
Johnson Matthey Plc	UK	92	1.34
Toyota Motor	JPN	91	0.87
Novartis	CHE	85	2.74
Hydro-Quebec	CAN	81	1.98
JEOL Ltd.	JPN	79	1.63
China Electronics Technology Group Corporation	CHN	79	0.82
Nippon Telegraph & Telephone	JPN	77	0.82
Bavarian Polymer Institute	DEU	73	1.34
State Grid	CHN	73	0.97
All-Russian Scientific Research Institute of Automatics	RUS	71	0.89
Johnson & Johnson	USA	70	1.69

Table A.1 Top 100 corporate entities by academic–corporate collaborative nano-publications (2015–19).—cont'd

Corporate name	Country code	Academic–corporate collaborative nano-publications	Field-weighted citation impact of academic–corporate collaborative nano-publications
Global Foundries, Inc.	USA	68	1.27
Western Digital	USA	67	1.17
Airbus Group	NLD	65	1.25
Laser Zentrum Hannover E.V.	DEU	64	1.97
UES, Inc.	USA	63	1.51
Merck	USA	58	4.89
China Aerospace Science and Industry Corporation	CHN	58	1.52
Toshiba	JPN	58	0.88
Oxford Instruments Group Plc	UK	56	1.55
SK Corporation	KOR	55	1.09
Hewlett-Packard	USA	54	4.36
China Electric Power Research Institute	CHN	54	0.95
F. Hoffmann-La Roche AG	CHE	53	4.87
General Electric	USA	52	1.67
TECNALIA	ESP	52	1.27
Federal State Unitary Enterprise	RUS	52	0.59
SABIC	SAU	49	2.02
The Energy Group	USA	48	2.23
China Minmetals Corporation	CHN	48	2.09
Bayer AG	DEU	48	2.08
Merck KGaA	DEU	48	1.44
Corning Incorporated	USA	48	1.23
RISE ICT	SWE	48	1.22
KeyW Corporation	USA	47	2.38

Continued

Table A.1 Top 100 corporate entities by academic–corporate collaborative nano-publications (2015–19).—cont'd

Corporate name	Country code	Academic–corporate collaborative nano-publications	Field-weighted citation impact of academic–corporate collaborative nano-publications
Total S.A.	FRA	47	1.11
Unilever	UK	46	1.09
Électricité de France S.A.	FRA	46	1.04
Microsoft USA	USA	45	3.51
Carl Zeiss SMT AG	DEU	45	1.80
Razi Vaccine and Serum Research Institute	IRN	45	1.01
Royal Dutch Shell PLC	NLD	44	1.73
Robert Bosch GmbH	DEU	44	1.52
Pfizer	USA	43	2.78
UCB S.A.	BEL	43	2.62
DuPont	USA	43	1.44
Dutch Polymer Institute	NLD	43	1.00
Schlumberger	USA	42	2.36
ExxonMobil	USA	42	1.77
DSM Food Specialties	NLD	42	1.60
ABB Group	CHE	42	0.74
TATA Steel	IND	41	1.37
Toray Industries Inc	JPN	41	1.15
Bruker BioSpin GmbH, Germany	DEU	40	1.23
3M	USA	39	1.75
Korea Electric Power	KOR	39	1.22
China National Offshore Oil Corp	CHN	39	1.14
CSEM SA	CHE	38	2.36
Haldor Topsoe AS	DNK	37	2.04
Siemens	DEU	37	1.27
Guangdong Power Grid Corporation	CHN	37	0.52

Table A.1 Top 100 corporate entities by academic–corporate collaborative nano-publications (2015–19).—cont'd

Corporate name	Country code	Academic–corporate collaborative nano-publications	Field-weighted citation impact of academic–corporate collaborative nano-publications
Rolls-Royce	UK	36	1.50
Northrop Grumman	USA	36	1.45
Panasonic Corporation	JPN	36	1.04
Procter and Gamble	USA	36	0.98
Novo Nordisk Foundation	DNK	35	1.83
Baoshan Iron and Steel Co. Ltd.	CHN	35	1.48
Micro Materials Limited	UK	35	1.38
AECOM	USA	35	1.23
Chevron Corporation	USA	35	1.22
Saint-Gobain S.A.	FRA	35	0.80
AbbVie	USA	34	2.67
Saudi Aramco	SAU	34	1.43

From Scopus.

APPENDIX

Methodology and indicators

B

Methodology

Our methodology is based on theoretical principles and best practices developed in the field of quantitative science and technology studies, particularly in science and technology indicators research. The *Handbook of Quantitative Science and Technology Research: The Use of Publication and Patent Statistics in Studies of S&T Systems* (Moed, Glänzel, and Schmoch, 2004)[1] gives a good overview of this field. It is based on the pioneering work of Derek de Solla Price (1978),[2] Eugene Garfield (1979),[3] and Francis Narin (1976)[4] in the United States, Christopher Freeman, Ben Martin, and John Irvine in the United Kingdom (1981, 1987),[5] and researchers in several European institutions including the Centre for Science and Technology Studies at Leiden University, the Netherlands, and the Library of the Academy of Sciences in Budapest, Hungary.

[1] Moed H., Glänzel W., & Schmoch U. (2004). *Handbook of Quantitative Science and Technology Research*, Kluwer: Dordrecht.

[2] de Solla Price, D.J. (1977–1978). "Foreword," *Essays of an Information Scientist*, Vol. 3, v–ix.

[3] Garfield, E. (1979). Is citation analysis a legitimate evaluation tool? *Scientometrics*, 1 (4), 359-375.

[4] Pinski, G., & Narin, F. (1976). Citation influence for journal aggregates of scientific publications: Theory with application to literature of physics. *Information Processing & Management* 12 (5): 297–312.

[5] Irvine, J., Martin, B. R., Abraham, J. & Peacock, T. (1987). Assessing basic research: Reappraisal and update of an evaluation of four radio astronomy observatories. *Research Policy*, 16(2-4), 213–227.

The analyses of bibliometric data in this book are based on recognized advanced indicators (e.g., the concept of field-weighted citation impact). Our base assumption is that such indicators are useful and valid, although imperfect and partial measures, in the sense that their numerical values are determined by research performance and related concepts, but also by other influencing factors that may cause systematic biases. In the past decade or so, the field of indicators research has developed best practices that state how indicator results should be interpreted and which influencing factors should be considered. Our methodology builds on these practices.

Identifying the relevant documents

For this book, a static version of the Scopus database covering 2000–2019 inclusive was aggregated by country, region, and subject. Subjects were defined by All Science Journal Classification subject areas (see Appendix D for more details).

- The search string[6] used to search nano-publications in Scopus is:

(TITLE-ABS-KEY (nano*) AND NOT TITLE-ABS-KEY (nano2 OR nano3 OR nano4 OR nano5 OR nanosecon* OR "NANO SECON*" OR "NANO GRAM*" OR nanogram* OR nanomol* OR nanophtalm* OR nanomeli* OR nanogeterotroph* OR nanoplankton* OR nanokelvin* OR nanocur)) AND DOCTYPE (ar OR re) AND PUBYEAR >1999 AND PUBYEAR <2020

Between 2000 and 2019, about 75,000 articles or publications in the field of "graphene" have been covered, and more than 54,000 articles have not been covered. "Two-dimensional materials" has 12,000 articles in the reporting period, of which 6,800 articles are not covered. Although the search string has some limitations in recalling all nano-related research outputs, based on the total recorded outputs of 1.4 million publications, the analysis we presented in the book is reliable and solid.

[6] Jiancheng Guan, Nan Ma, China's emerging presence in nanoscience and nanotechnology: A comparative bibliometric study of several nanoscience 'giants', Research Policy, Volume 36, Issue 6,2007, Pages 880–886, ISSN 0048-7333.

Publication types used in the analysis

Throughout this book, analyses only include these publication types that are indexed in Scopus:

- Articles
- Reviews

Counting

All analyses use whole counting rather than fractional counting. For example, if a paper has been coauthored by one author from China and one author from the United States, that paper counts toward the publication count of China as well as the publication count of the United States. Total counts for each country are the unique count of publications.

Bibliometric indicators

Author refers to any individual listed in the author byline of a Scopus-indexed publication.

Citation is a formal reference to earlier work made in an article or patent, frequently to other journal articles. A citation is used to credit the originator of an idea or finding and is usually employed to indicate that the earlier work supports the claims of the work citing it. The number of citations received by an article from subsequently published articles is a proxy of the importance of the reported research.

Cross-sector collaboration (academic–corporate/academic–medical collaboration) is defined as a publication in which either *corporate* or *medical* entities are included in the author affiliation byline in addition to authors affiliated to academic entities. Academic–corporate and academic–medical collaboration are terms used in this book to denote, respectively, the count of collaborations (i.e., research collaborations) between authors from academic and corporate sectors and between authors from academic and medical sectors.

Compound annual growth rate is defined as the year-over-year constant growth rate over a specified period. Starting with the first value in any series and applying this rate for each of the time intervals yields the amount in the final value of the series.

The following definition holds:

$$\mathrm{CAGR}(t_0, t_1) = \left(\frac{V(t_1)}{V(t_0)}\right)^{\frac{1}{t_1 - t_0}} - 1$$

where $V(t_0)$ is the start value and $V(t_1)$ is the final value of a time series and difference $t_1 - t_0$ defines the length of the time interval.

Field-weighted citation impact[7] is an indicator of mean citation impact and compares the actual number of citations received by an article with the expected number of citations for articles of the same document type (article, review, or conference proceeding paper), publication year, and subject field. When an article is classified in two or more subject fields, the harmonic mean of the actual and expected citation rates is used. The indicator is therefore always defined with reference to a global baseline of 1.0 and intrinsically accounts for differences in citation accrual over time, differences in citation rates for different document types (reviews typically attract more citations than research articles, for example), as well as subject-specific differences in citation frequencies overall and over time and document types. It is one of the most sophisticated indicators in the modern bibliometric toolkit (see also **Citation**).

International collaboration (i.e., research collaboration) in this book is indicated by articles with at least two different countries listed in the authorship byline.

Institutional collaboration (i.e., research collaboration) in this book is indicated by articles with a single institute listed in the authorship byline.

---

[7] Amrita Purkayastha, Eleonora Palmaro, Holly J. Falk-Krzesinski, Jeroen Baas, Comparison of two article-level, field-independent citation metrics: Field-Weighted Citation Impact (FWCI) and Relative Citation Ratio (RCR), Journal of Informetrics, Volume 13, Issue 2, 2019, Pages 635–642, ISSN 1751-577, https://doi.org/10.1016/j.joi.2019.03.012.

Output/publication output for an institute or country is the count of articles with at least one author from that institution or country, respectively (according to the affiliation listed in the authorship byline). All analyses use both full and fractional counting. A **full publication** is a publication to which at least one author affiliated with the institution has contributed.

National collaboration (i.e., research collaboration) in this book is indicated by articles with at least two different institutes from the same country listed in the authorship byline.

Single-author publication is a publication in which only a single author is listed in the authorship byline.

Top 1% highly cited publications are those among the top 1% based on citations of all articles published and cited in a given period. An institution's number or share of highly cited articles is treated as indicative of the excellence of their research.

Topics (as pertaining to Topics of Prominence), refer to nearly 96,000 research topics created using citation patterns of Scopus-indexed publications. The methodology for using citation patterns to define research topics was developed through an Elsevier collaboration with research partners. The advantage of taking a citation-based approach to identify research topics is that one need not rely on identifying all relevant keywords to define a research area. Rather, the research area is delineated by citation patterns in the topic, in which research that appears in the same citation network is clustered together in the same topic. This approach provides a more nuanced definition of the research topic.

Topic clusters are a higher-level aggregation of these research topics based on the same direct citation algorithm that creates the topics. Although topics are easy to understand for the subject experts, they are more difficult for subject generalists. To aid discovery and understanding of the topics, we have taken the topics and aggregated them to around 1500 topic clusters. When the strength of the citation links between topics reaches a threshold, a topic cluster is formed. More details on topics are available at https://service.elsevier.com/app/answers/detail/a_id/28428/.

The **Prominence score** is an indication of the momentum related to a particular topic. The Prominence score is calculated by considering three metrics:

- Citation Count in year n to papers published in n and $n-1$
- Scopus Views Count in year n to papers published in n and $n-1$
- Average CiteScore for year n

The equation for Prominence score for each topic j in year n is:

$$P_j = \frac{0.495\left(C_j - \mathrm{mean}\left(C_j\right)\right)}{\mathrm{stdev}\left(C_j\right)} + \frac{0.391\left(V_j - \mathrm{mean}\left(V_j\right)\right)}{\mathrm{stdev}\left(V_j\right)} + \frac{0.114\left(CS_j - \mathrm{mean}\left(CS_j\right)\right)}{\mathrm{stdev}\left(CS_j\right)}$$

Values that appear in this equation are log-transformations of the raw values in which:

$$C_j = \ln\left(c_j + 1\right)$$
$$V_j = \ln\left(v_j + 1\right)$$
$$CS_j = \ln\left(cs_j + 1\right)$$

C_j = citations in year n of publications in cluster J published in year n and $n-1$

V_j = views in year n of publications in cluster J published in year n and $n-1$

CS_j = weight average Citescore of journals in year n containing publications in cluster J.

Prominence, as defined here, is a linear combination of citations, views, and journal impact for a given topic, in which each factor is normalized by the topic standard deviation.[8]

Topic Prominence score has been shown to correlate with topic level funding per author in a US sample of funded grants. The analysis found that on average, the higher the Prominence score, the more money per US author was available for research on that topic.

---

[8] https://www.sciencedirect.com/science/article/pii/S1751157717302110.

Patent indicators

Technology Relevance is based on forward citations. Technology Relevance measures whether a patent has been cited more often than have other patents from the same technology field and year, while also considering that international patent offices follow different citation rules. The total number of patent citations received depends on the technology field of the patented invention and also the time that has passed since the patent was published. Patents only recently published tend to have received much fewer citations than older patents. The time-dependency of citations is corrected by dividing the number of citations received by a patent by the average number of citations received by all patents published in the same year.

Technology Relevance also considers that international patent offices follow different citation rules. Therefore, the number of patent citations is corrected for age, patent office citation practice, and technology field. It is a relative measure comparing one patent with other patents. A value of two means that the patent is twice as relevant for subsequent developments as an average patent in the same technology field and of the same age.

Market Coverage: The total size of the worldwide markets in which patent protection exists. The more patents a patentee (in this case an institution or a country the patent owner is affiliated with) owns in important markets, the more valuable the patents are estimated to be. This is because innovators spend more effort and resources on protection in multiple (global) markets via patents if they believe an invention is more valuable.

Market Coverage is calculated based on granted and pending patents; hence, valid patents per country are adjusted for each market's size, as opposed to simple country counts. The size of each market is estimated using the sum of countries' gross national income (GNI) relative to the US GNI as the largest global economy. A Market Coverage of 2 means that the protected markets are, in total, twice as large as the US market alone.

Competitive Impact: The economic value of a patent as measured by its Technology Relevance and Market Coverage. Competitive Impact is stated relative to the other patents in the same field (e.g., a value of 3 means that the patent is three times as important as the average patent in the field).

Patent Asset Index: The Patent Asset Index[9] of a patent portfolio is defined as the aggregate strength of all patents the portfolio contains. The strength of each individual patent is measured by its Competitive Impact. The PatentSight Competitive Impact consists of two dimensions: PatentSight Technology Relevance and PatentSight Market Coverage.

[9] Ernst, H., & Omland, N. (2011). The patent asset index - A new approach to benchmark patent portfolios. World Patent Information, 33(1), 34-41. https://doi.org/10.1016/j.wpi.2010.08.008.

Data sources

Scopus

Scopus is Elsevier's abstract and citation database of peer-reviewed literature, covering 77.3 million documents published in over 39,000 journals, book series, and conference proceedings by some 5,000 publishers.

Scopus coverage is multilingual and global: approximately 46% of titles in Scopus are published in languages other than English (or published in both English and another language). In addition, more than half of Scopus content originates from outside North America, representing many countries in Europe, Latin America, Africa, and the Asia Pacific region.

Scopus coverage is also inclusive across all major research fields, with 13,300 titles in the physical sciences, 14,500 in the health sciences, 7,300 in the life sciences, and 12,500 in the social sciences (the latter including some 4,000 arts and humanities–related titles). Titles covered are predominantly serial publications (journals, trade journals, book series, and conference material), but considerable numbers of conference papers are also covered from stand-alone proceedings volumes (a major dissemination mechanism, particularly in the computer sciences). Acknowledging that a great deal of important literature in all fields (especially in the social sciences as well as the arts and humanities) is published in books, Scopus has begun to increase book coverage in 2013. As of 2018, Scopus includes 1.75 million book items, 400,000 of which are in the social sciences and 290,000 of which are in the arts and humanities.

More information can be found at: https://www.scopus.com/

SciVal

Elsevier's new generation of SciVal offers quick, easy access to the research performance of over 12,000 research institutions and 230 nations worldwide. A ready-to-use solution with unparalleled power and flexibility, SciVal enables you to navigate the world of research and devise an optimal plan to drive and analyze your performance.

SciVal builds on Elsevier's extensive experience over several years, working with many leading institutions worldwide via SciVal Spotlight and SciVal Strata.

Data source: SciVal is based on output and usage data from Scopus, the world's largest abstract and citation database for peer-reviewed publications.

SciVal uses Scopus data from 1996 to the present, covering over 48 million records in 21,000 serials from 5,000 publishers. These include:

- more than 22,000 peer-reviewed journals
- 360 trade publications
- 1,100 book series
- 5.5 million conference papers

SciVal offers a broad spectrum of industry-accepted and easy-to-interpret metrics including Snowball Metrics, which are global standard metrics defined and agreed upon by higher education institutions for institutional strategic decision-making through benchmarking. Metrics in SciVal help institutions measure their or a country's productivity, citation impact, collaboration, subject discipline, and more.

More information can be found at: https://www.scival.com/

PatentSight

PatentSight provides unique, reliable, and relevant insights into the patent landscape for decision-makers and patent experts in the fields of benchmarking, research and development strategy, trend-scouting, merger and acquisition, licensing, and patent portfolio optimization.

It provides simple measures that can help assess the breadth, strength, and commercial importance of a patent or patent family and yield insights into potential commercial importance. It also allows for geographic and company-based analysis of patent portfolios to help judge market penetration and readiness on a country-by-country basis.

The patent databases INPADOC and DOCDB provided by the European Patent Office (EPO) are the central patent data sources for the PatentSight database. Furthermore, we add legal status information from national offices (United States and Japan) to the data stream, as well as subsidiary information from corporate structure databases. In addition, all patent offices around the world provide patent data with the EPO, so we cover all relevant patent data worldwide. The PatentSight database currently covers patent data from more than 80 patent authorities.

Because PatentSight is part of LexisNexis, our database also contains IP DataDirect (IPDD). IPDD from LexisNexis provides worldwide patent information in bulk. The entire IPDD content with improved bibliographic and legal status information will be integrated into the PatentSight database by the end of 2021.

More information can be found at: https://www.patentsight.com/en/

Elsevier funding institutional

This platform offers a competitive edge, providing a holistic view of the research funding landscape and using a single solution that combines over 18,000 active funding opportunities with information on over three million awarded research grants from a wide range of funders.

Funding Institutional offers access to the following content:

- more than 3500 government and private funding organizations, including 2000 funders in the United States
- over 18,000 active funding opportunities

- greater than US $98 billion worth of active funding opportunity
- more than 3 million awarded grants
- greater than US $1.6 trillion worth of awarded grants

More information can be found at: https://www.elsevier.com/solutions/funding-institutional

All science journal classification subjects

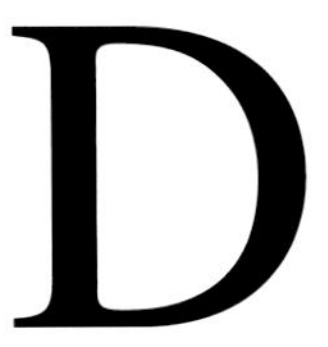

Titles in Scopus are classified under four broad subject areas (life sciences, physical sciences, health sciences and social sciences and humanities), which are further divided into 27 subjects and more than 300 fields. Titles may belong to more than one subject area.

Scopus 27-subject classification	Subjects used in knowledge flow diagram of nanoscience	Scopus 27-subject classification	Subjects used in knowledge flow diagram of nanoscience
Agricultural and biological sciences	Others	Health professions	Health science
Arts and humanities	Social science	Immunology and microbiology	Others
Biochemistry, genetics, and molecular biology	Biochemistry, genetics, and molecular biology	Materials science	Materials science
Business, management, and accounting	Social science	Mathematics	Others
Chemical engineering	Chemical engineering and chemistry	Medicine	Health science
Chemistry	Chemical engineering and chemistry	Multidisciplinary	Others
Computer science	Others	Neuroscience	Others
Decision sciences	Social science	Nursing	Health science
Dentistry	Health science	Pharmacology, toxicology, and pharmaceutics	Pharmacology, toxicology, and pharmaceutics

Continued

Scopus 27-subject classification	Subjects used in knowledge flow diagram of nanoscience	Scopus 27-subject classification	Subjects used in knowledge flow diagram of nanoscience
Earth and planetary sciences	Others	Physics and astronomy	Physics and astronomy
Economics, econometrics, and finance	Social science	Psychology	Social science
Energy	Energy	Social sciences	Social science
Engineering Environmental science	Engineering Environmental science	Veterinary	Health science

Country codes

Code	Country	Code	Country
AUS	Australia	KOR	South Korea
BEL	Belgium	NLD	Netherlands
CAN	Canada	NOR	Norway
CHE	Switzerland	POL	Poland
CHN	China	RUS	Russia
DEU	Germany	SGP	Singapore
ESP	Spain	SAU	Saudi Arabia
FIN	Finland	SWE	Sweden
FRA	France	JPN	Japan
IND	India	UK	United Kingdom
ITA	Italy	USA	United States

Index

Note: 'Page numbers followed by "f" indicate figures and "t" indicate tables.'

Q

R

S

T

U

V

W

Printed in the United States
by Baker & Taylor Publisher Services